Temperature

1 K	= 8.62×10^{-5} eV
1 eV	= 1.1605×10^4 K
1 GeV	= 1.605×10^{13} K

Fundamental Constants

Gravitation	G	= 6.673×10^{-8} cm³/g sec²
Speed of light	c	= 2.997×10^{10} cm/sec.
Planck's constant	h	= 6.6261×10^{-27} cm² g/sec
Planck's reduced constant	$\hbar$	= 1.054592×10g cm²/sec
		= 2.612×10^{-66} cm²
Electromagnetic energy constant ($E = aT^4$)	a	= 7.5647×10^{-15} erg/cm³ K⁴
Boltzmann's constant	k	= 1.38×10^{-16} erg/K
		= 8.617×10^{-5} eV/K
		= 1.38×10^{-16} g cm²/sec² K
Elementary electrical charge	e	= 1.6022×10^{-19} C
Proton rest mass	m_p	= 1.673×10^{-27} kg
Electron rest mass	m_e	= 9.1×10^{-31} kg
Fermi constant	G_F	= 1.664×10^{-5} GeV⁻²
		= $(292.8 \text{ GeV})^{-2}$
Planck length	l_{PL}	= 1.61×10^{-33} cm
mass	M_{PL}	= 2.17×10^{-5} g
energy	E_{PL}	= 1.2211×10^{19} GeV
time	t_{PL}	= 5.39×10^{-44} g/cm³
density	d_{PL}	= 5.158×10^{93} g/cm³
Bohr radius	a_0	= 5.29×10^{-9} cm
Electron radius	r_e	= 2.818×10^{-13} cm

The Multi-Universe Cosmos

The First Complete Story of the Origin of the Universe

The Multi-Universe Cosmos

The First Complete Story of the Origin of the Universe

A. Karel Velan
Velan, Inc.
Montreal, Quebec
Canada

Plenum Press • New York and London

```
Library of Congress Cataloging-in-Publication Data

Velan, A. Karel.
    The multi-universe cosmos : the first complete story of the origin
  of the universe / A. Karel Velan.
       p.   cm.
    Includes bibliographical references and index.
    ISBN 0-306-44267-1
    1. Cosmology--Popular works. 2. Astrophysics--Popular works.
  I. Title.
  QB982.V45  1992
  523.1--dc20                                                   92-22568
                                                                     CIP
```

Cover: An ocean of stars with interstellar dust and gas called the North America Nebula in Cygnus. (Courtesy Astrophoto Laboratory)

ISBN 0-306-44267-1

©1992 Plenum Press, New York
A Division of Plenum Publishing Corporation
233 Spring Street, New York, N.Y. 10013

All rights reserved

No part of this book may be reproduced, stored in a retrieval system, or transmitted in any form or by any means, electronic, mechanical, photocopying, microfilming, recording, or otherwise, without written permission from the Publisher

Printed in the United States of America

Preface

In light of the barrage of popular books on physics and cosmology, one may question the need for another.

Here, two books especially come to mind: Steven Weinberg's *The First Three Minutes*, written 12 years ago, and the recent best-seller *A Brief History of Time* by Stephen Hawking. The two books are complementary. Weinberg—Nobel prize winner/physicist—wrote from the standpoint of an elementary particle physicist with emphasis on the contents of the universe, whereas Hawking wrote more as a general relativist with emphasis on gravity and the geometry of the universe. Neither one, however, presented the complete story. Weinberg did not venture back beyond the time when temperature was higher than 10^{13} K and perhaps as high as 10^{32} K. He gave no explanation for the origin of particles and the singularity or source of the overwhelming radiation energy in our universe of one billion photons for each proton. Hawking presents a universe that has no boundaries, was not created, and will not be destroyed.

The object of this book is to describe my new theory on the creation of our universe in a multi-universe cosmos. The new cosmological model eliminates the troublesome singularity–big bang theory and explains for the first time the origin of matter and the overwhelming electromagnetic radiation contained in the universe. My new theory also predicted the existence of high-energy gamma rays, which were recently detected in powerful bursts. There is no known process in the universe that can produce these powerful gamma rays at energy levels of 10^{11} GeV and more.

The book is intended for readers with a basic scientific education and those interested in astronomy and physics. The book contains complicated scientific ideas, but I employ simple explanations and mathematics to substantiate the arguments. Any reader of *A Brief History of Time* by Stephen Hawking will appreciate reading my book.

I am extremely grateful for the advice, support, encouragement, and the reading of my manuscript by Sir Martin Rees of Cambridge and Professor Ken Pounds of Leicester University.

I am particularly grateful to Professor Dr. Raine of Leicester for his careful editing of the manuscript and his many valuable suggestions and for the patience of my secretary Bea Dow.

A. Karel Velan

Montreal

Contents

1. Cosmology—Particles and Forces 1
2. Properties of Elementary Particles, the Building Blocks of the Universe 7
3. Basic Principles of Quantum Mechanics 21
4. Einstein's Relativistic Properties of Particles 35
5. Quantum Electrodynamics 41
6. Quantum Chromodynamics, the Strong Nuclear Force 51
7. Conservation Laws 61
8. The Particle Zoo 71
9. The Weak Nuclear Force 83
10. Gravitation 91
11. Black Holes, Quasars 99
12. Unification of the Four Forces 113
13. Review of Modern Cosmological Theories 121
14. Vacuum of Space 161
15. The Velan Multi-Universe Cosmos, the First Cosmological Model without Singularity 179
16. The Primordial Electromagnetic Radiation Field: Postulate 6 187
17. The Creation Process 193
18. The Velan Fireball 207

19. The Big Bang of the Fireball	219
20. The History of Evolution of the Early Universe	225
21. Galaxy Formation	245
22. The Birth of the First Generation of Stars	267
23. The Fate of the Sun and Planetary System	279
24. The Death of Stars	293
25. The Fate of the Universe	329
Epilogue	345
Glossary	353
Index	361

Color plates on the following four pages:

Page ix: An ocean of stars with interstellar dust and gas called the North America Nebula in Cygnus. (Astrophoto Laboratory)

Page x: The Whirlpool galaxy (M51) of over 100 billion stars, 35 million light-years from earth. (U.S. Naval Observatory)

Page xi: The Orion Nebula, a galactic incubator where new stars are born. (U.S. Naval Observatory)

Page xii: The Pleiades, a cluster of young stars 410 light-years away. (Astrophoto Laboratory)

1

Cosmology—Particles and Forces

It is man's natural curiosity that for centuries has initiated the quest to better understand the composition and evolution of our universe, the structure of its smallest particles of matter, and the fundamental laws that govern the universe, which may one day come to its end. As long ago as 2000 years, the Greek philosopher Democritus claimed that matter must be composed of different types of small grains.

The science dealing with the birth and evolution of the universe is called cosmology. To understand cosmology, the history of creation and evolution of the universe, is to understand well the physics of elementary and nuclear particles, their interactions and behavior under different levels of energy fields, density, and temperature. It is equally important to get a proper perspective on the various laws governing the behavior of particles and the four forces of nature controlling all events in the universe.

If we could break down all matter in the universe into elementary, indivisible particles, we would find two types of quarks, u and d, neutrinos, and electrons. The quarks, which were born free, according to my theory, combined later into nuclear particles named protons and neutrons, which, together with electrons, formed atoms.

Nevertheless, in high-energy particle accelerators in many laboratories throughout the world, such as CERN in Geneva, hundreds of other nuclear particles and antiparticles (antimatter) are being created from concentrated electromagnetic energy when high-speed particles collide and annihilate into electromagnetic energy or γ rays. These particles, however, also created in the upper atmosphere by collisions with cosmic radiation, are all short-lived, some existing for only 10^{-23} second, and decay into the four stable particles—protons, neutrons, neutrinos, and electrons. Free neutrons, not confined in atoms, also decay into protons, electrons, and electron neutrinos. It would appear, from these experiments, that particles of matter are just another form of energy.

The four elementary particles, three of which are contained in protons and neutrons, constitute all known matter in the universe, including galaxies, stars, and our own bodies. Atoms consist of a core of protons and neutrons and orbiting electrons. The four elementary, stable particles, which are the building blocks of the universe, can be put into two groups:

1. Electrons e^- and electron-type neutrinos ν_e, which exist independently and are not all confined to atoms.
2. Quarks u (up) and d (down), which automatically group together in threes, under the influence of the strong nuclear force and create nuclear particles such as protons and neutrons. They can also combine with an antiquark to form short-lived mesons. All quarks are confined in protons and neutrons; to date, no free quarks have been found in the universe.

The four basic, elementary particles are pointlike and extremely small in size. The quarks u and d have an approximate size of one-third of a proton or $\frac{1}{3} \times 10^{-13}$ cm. The electron has a radius of approximately 10^{-17} cm. The neutrino is probably massless, although this has not been determined. If the neutrino has mass, it would, at best, be pointlike to the same limit as an electron. These fundamental particles are difficult to visualize. It is amazing that these dots and points were enough to create the enormous variety of things and beings that exist on earth and in the universe, and that the particles never stopped combining and recombining to form objects and living beings of infinite variety. According to my theory, these particles, which were created by transformation from a powerful, cosmic radiating energy 18 billion years ago, interacted, stabilized, and together with electromagnetic photons, formed the universe as we know it today.

In addition to the particles of matter, the universe contains an enormous quantity of photons and free neutrinos. Photons are massless particles or quanta of electromagnetic radiation. For each proton in the universe, there are 1 billion (10^9) photons and approximately the same number of neutrinos. The origin of these massive, electromagnetic photons is also explained for the first time in my new theory.

The interactions between particles of matter are affected by four forces of nature, namely the strong nuclear force, the weak nuclear force, the electromagnetic force, and gravitation. This is explained, schematically, in Figure 1.1.

1. The strong nuclear force, which is the most powerful, binds together elementary quarks in protons and neutrons. The strong force also binds protons with neutrons into a nucleus of atoms.

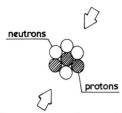 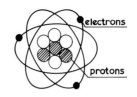

1. The strong force binds the nucleus of an atom.
2. The electromagnetic force binds electrons to protons in atoms.

3. The weak force radioactive decay of a neutron
neutron → proton + electron + electron neutrino
4. The gravitational force keeps earth in orbit and curves the space around the sun.

Figure 1.1. Interactions of particles and the four forces of nature.

2. The electromagnetic force, which acts between particles, carrying an electric charge, binds clouds of electrons ($-$) with protons ($+$) in the nucleus of atoms. It is 1000 times weaker than the strong force. As an example, the simplest atom—hydrogen—consists of one proton and one circling electron, held together by the electromagnetic force.
3. The weak nuclear force influences the decay of particles when for instance a neutron in the nucleus of an atom or a free neutron with a half-life of 15 minutes breaks up and releases an electron, proton, and electron-neutrino. It is enormously weaker than the strong force. Its relative intensity is 10^{-24} when the strong force is set equal to unity (1).

    The weak and electromagnetic forces are related and their unified forces at higher energies, larger than 100 GeV, are known as the "electroweak" force.
4. The gravitational force acts between particles and pulls all matter together. It controls the movements of planets around the sun or of stars around the center of a galaxy, and is responsible for the collapse of large stars into black holes.

Figure 1.2. On this screen at CERN you see particle tracks from a bubble chamber detector. Electrons and positrons spiral in opposite directions in the chamber's magnetic field. (CERN)

We can now commence a detailed study of particle physics and their governing laws, starting with the elementary particles, the building blocks of matter.

Modern particle physics, confirmed by many experiments, demonstrate that we must look at matter as a compact, amazingly concentrated form of energy. The process of transformation of energy into matter cannot be observed in everyday life for three main reasons.

1. Under normal conditions, energy is not sufficiently concentrated to produce particles of matter.
2. Particles created in modern accelerators–colliders from concentrated energy, released by collision of elementary particles accelerated to near speed-of-light velocities, are so tiny that they are invisible to the naked eye.
3. In addition, these particles have an enormously short life, many times only 10^{-23} of a second. Fortunately, however, they leave traces during their short flight in special chambers of saturated vapor and electromagnetic fields. They can be photographed and their paths reveal their characteristics before they finally disintegrate into known stable particles (Figure 1.2).

All matter, as we know it, is composed of three stable nuclear particles—the proton, neutron, and electron, though a free neutron is also unstable and disintegrates on its own in approximately 15 minutes into a proton, electron, and neutrino. A neutron is stable only in cores of atoms. Protons and neutrons, as we have already mentioned, are composed of three quarks. Only two types of quarks in a combination of three form a proton or neutron. They are the up or down quarks, u and d.

To the best of our knowledge, no free quarks can be found in the universe. They are all part of protons, neutrons, mesons, and many other variations of similar short-lived particles created in accelerators, and in collision of cosmic radiation with particles in the earth's upper atmosphere. Free quarks existed, however, in the early universe when energies were extremely high, ranging from 10^{15} GeV to 10^{19} GeV and difficult-to-imagine temperatures of up to 10^{27} K. (1 GeV equals 10^9 eV or 1 billion electron volts.)

At this time the quarks were compressed to high densities and at such small distances below 10^{-13} cm that the strong nuclear force, which was anyway substantially weaker and which interacts by means of gluons, could not cause the confinement of quarks into protons. The quarks enjoyed freedom for a short time.

The first proof that energy can transform into matter came during the 1930s when "cosmic rays" from outer space were observed to be hitting particles in the outer atmosphere, transforming them into new, short-lived particles called mesons, consisting of one quark and one antiquark.

Nevertheless, already in 1905, Einstein, with his unique ability to unveil secret laws of nature without the benefit of experimental investigations, predicted that energy behaved as if it had mass and that mass of matter can be regarded as condensed energy.

He expressed this neatly and precisely in his famous formula, later proven to be accurate, $E = mc^2$. This means, as an example, that 25 million kilowatt-hours "weigh" 1 gram and that in order to create 1 gram of matter, it would be necessary to condense 25 million kilowatt-hours of energy into an incredibly small volume of space. This amount of energy would be equivalent to the entire electrical power used in a day by very large cities.

In particle physics, one of the units of energy used is the gigaelectron volt (GeV), i.e., one billion electron volts. It is approximately the amount of energy that must be condensed to form one proton. To be exact, the mass-energy of a proton is 0.9382 GeV and the mass-energy of an electron only 0.000511 GeV. (One electron volt is equal to 1.602×10^{-12} erg, or 1.602×10^{-19} joule, 1.782×10^{-33} g, 1 GeV = 10^9 eV).

The largest present-day accelerators endow protons with energies of up to 600 GeV. This sounds like an enormous amount of energy but only because it is concentrated into an extremely small particle. In contrast to the 600 GeV, energy levels of 10^{15} GeV to 10^{20} GeV prevailed during the creation of the universe.

Two flying mosquitoes, which carry much more energy, create, when colliding, no new particles. Two protons, however, colliding at these energies, produce new particles from the kinetic energy released during the collision.

It would probably take an accelerator–collider the size of our Milky Way galaxy to re-create the unique high-energy-particle "laboratory" provided by nature during the creation period.

2

Properties of Elementary Particles, the Building Blocks of the Universe

2.1. ELECTRONS

The electron is the lightest elementary particle, discounting neutrinos. All chemical properties of atoms and molecules are influenced and determined by the electric interaction of electrons with each other and atomic nuclei.

The electron is a pointlike, stable particle with no substructure. The classical electron radius can be calculated from the formula

$$r_e = \frac{e^2}{m_e c^2} = 2.81 \times 10^{-17} \text{ cm}$$

where e is the electric charge, m_e is the mass of the electron, and c is the speed of light.

The electron does not interact with the strong nuclear force but interacts with the weak and electromagnetic force and, therefore, belongs to the group called leptons.

The electron has four properties:

1. *Mass m_e.* The mass of an electron expressed in energy units of 0.00051 GeV is approximately 2000 times smaller than that of a proton. The mass expressed in grams is only 0.9×10^{-27} g, in ergs 8.187×10^{-7} erg, in degrees K 5.93×10^9 K, and in centimeters 1.24×10^{-52} cm.

2. *Electrical charge Q.* The electron has an electrical charge of -1. The amount of the charge is equal to the charge of a proton but it is negative. The antiparticle of an electron, called a positron, has the same amount of electrical charge but it is positive ($+1$), the same as a proton.

3. *Spin "J."* The electron rotates around its own axis or has its own angular

momentum, and in an atom it rotates around the nucleus. In the hydrogen atom, the single electron moves in an orbit of lowest possible energy around the nucleus, which in this case consists of one proton (Figure 2.1). If an electron changes its spin orientation in an orbit of the same radius, it emits a photon due to the difference in energy of the two spins.

If additional energy is supplied to the electron from an outside source, it will move to a higher orbit (dashed line in Figure 2.1) but only for less than 10^{-8} of a second, whereupon it falls back to the ground state and emits the extra energy in the form of electromagnetic radiation (light in a neon bulb). The spin or speed at which the electron rotates has an influence on the energy associated with its orbit.

We measure spin in units of the Planck constant $\hbar$. In terms of Planck's quantum theory, which will be explained later,

$$\hbar = \frac{h}{2\pi} \qquad (2.1)$$

The various angular momenta can be 0, 1, 2, 3, ... The angular momentum is a vector. It has direction and magnitude.

Planck's constant $\hbar$ is set at 1; therefore, the spin of the electron becomes $+\frac{1}{2}$. The component along any given axis can be $+\frac{1}{2}$ or $-\frac{1}{2}$. The axis of the spin rotation is perpendicular to the particle momentum and the spin does not change

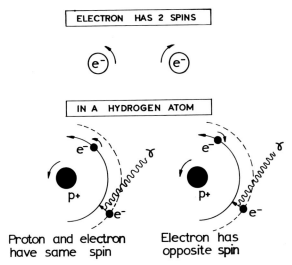

Figure 2.1. Origin of 21-cm emission line of neutral hydrogen. When an electron "changes its spin" from the same direction as a proton, a radio photon of wavelength 21 cm is emitted. If the electron moves to a higher energy orbit due to an outside energy source, it returns to the ground state orbit and also emits a photon or γ ray.

Properties of Elementary Particles

in free motion. The electron has two possible motions as shown by the arrows in Figure 2.1.

An electron in an atom has four quantum numbers: energy, angular momentum, the 2-components of the total angular momentum, and spin. Only one electron can have a given set of quantum numbers and rotate in the same orbit.

4. *Magnetic moment.* As the electrical charge of the electron moves around in a circle, similar to a circulating electric current, it generates a small magnetic field and force (Figure 2.2). The magnetic moment μ can be calculated from the formula

$$\mu = \frac{h}{m_e} \qquad (2.2)$$

The relationship between the four properties—magnetic moment μ, electrical charge Q, mass m_e, and spin J—can be expressed as follows:

$$\mu = -g \frac{Q}{4\pi m_e c} J \qquad (2.3)$$

where c = speed of light g = constant of proportionality or g factor which is a dimensionless number, m_e = mass, and Q = electrical charge. The minus sign of g indicates that the vector of the magnetic moment ($-$) is antiparallel to the spin vector J (Figures 2.3 and 2.4).

Theoretical calculations have established that g has a value of 2. Calculations based on extensive experiments to determine the behavior of a single electron trapped in a large magnetic device by P. Ekstrom and D. Wineland have accurately determined the factor g to be 2.0023193044.

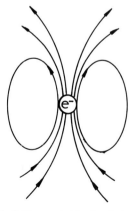

Figure 2.2. Magnetic field generated by the electron's electrical charge.

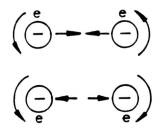

Figure 2.3. Two electrons with the same direction of spin will attract themselves magnetically. They will repulse if their spins are opposite.

We can therefore write the equation as

$$\mu = -2.0023193044 \frac{Q}{4\pi mc} J \tag{2.4}$$

As all properties of the electron such as charge Q and mass m as well as the speed of light c are known, the magnetic (μ) vector moment can be calculated when the angular spin moment (J) is known.

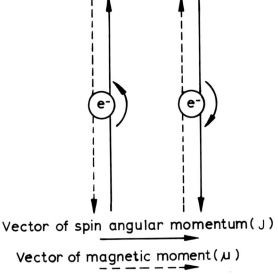

Figure 2.4. Solid arrow, vector of spin angular momentum (J); dashed arrow, vector of magnetic moment (μ).

2.2. NEUTRINOS (ν)

The neutrino is a massless or nearly massless, electrically neutral, elementary particle. Its mass has not been determined but does not exceed 30 eV, i.e., or 10,000 times lighter than an electron. Particle institutes that have tried to measure the neutrino mass have reported the following: Los Alamos, 27 eV; Munich, 15 eV; Moscow, 26 eV: Tokyo 28 eV; and CERN, 18 eV.

The neutrino participates only in weak and gravitational interactions and comes in three variations:

- Electron-neutrino (ν_e), the lightest, perhaps 30 eV
- Muon-neutrino (ν_μ), perhaps 285 keV
- Tau-neutrino (ν_τ), the heaviest, perhaps 35 MeV

There are three antineutrino equivalents: $\bar{\nu}_e$, $\bar{\nu}_\mu$, and $\bar{\nu}_\tau$.

The neutrino was discovered during an analysis of neutron decays by Pauli. The products of neutron decay, also called β-decay, are a proton, electron, and electron-antineutrino:

$$n \rightarrow p + e^- + \bar{\nu}_e$$

The antineutrino accounts for the missing energy as a neutron is slightly more massive than a proton. Based on the conservation law of spin, the neutrino has a spin of $\frac{1}{2}$. This can also be determined from the neutron decay formula:

$$\text{Spin } \tfrac{1}{2} \rightarrow \tfrac{1}{2} - \tfrac{1}{2} + \tfrac{1}{2} = \tfrac{1}{2}$$
$$n \rightarrow p + e^- + \bar{\nu}_e$$

Neutrinos ensured that the law of conservation of energy and momentum was maintained in the β-decay of neutrons.

Neutrinos interact only very weakly with matter and can penetrate large masses such as the earth without being disturbed. Millions upon millions of neutrinos are produced in nuclear reactors from the radioactive decay of reactor materials and the decay of free neutrons.

Large underground detectors are useful in capturing neutrino fluxes from the sun and supernova explosions. The issue of the mass of neutrinos is very important in astrophysics. If neutrinos are massive, a large part of the total mass in the universe would consist of neutrinos. Neutrinos belong to the lepton family. Both the electron and the neutrino have baryon number 0, which we discuss later.

2.3. u (UP) AND d (DOWN) QUARKS

Quarks u and d are more massive than electrons. They are pointlike, elementary particles of matter now confined in protons, neutrons, mesons, and many other unstable nuclear, quark compound particles. No free quarks exist, to our knowledge, at the energy and temperature levels prevailing in the universe today. However, they were born as free particles from powerful energy at the time of creation and maintained their free status until the young universe expanded and energy levels dropped below 10^{15} GeV.

It has been proven in experiments at high energies much below the threshold of 10^{12} GeV, which obviously cannot be re-created in a laboratory, that quarks behave as free particles at very high energies, a phenomenon called asymptotical freedom.

At the time of creation, not only were the energies high (above 10^{15} GeV), but the distances between particles, including quarks, were very small, less than 10^{-14} cm. At these small distances, the attractive forces between quarks were negligibly small and the quarks behaved as free particles at a relative moment p of

$$\Delta p > \frac{1 \text{ GeV}}{c} \qquad (2.5)$$

where c is the speed of light.

The potential energy of quark-to-quark interaction rises infinitely with distance, which is called *infrared confinement*. This increase is so rapid that two quarks cannot be separated beyond the radius of a proton, i.e., 10^{-13} cm.

Quarks have four basic characteristics:

1. *Mass m_q.* The mass of a quark u or d has not been determined experimentally; however, it is estimated to be one-third of the mass of the proton or 0.3127 GeV. Other types of quarks, as will be seen later, are much heavier.

2. *Electrical charge "Q."* Quarks have an electrical charge. The elementary charge is -1 for an electron, which is equivalent in magnitude to the charge of a proton ($+1$). The electrical charges of quarks u and d are nonintegral. The u quark has a charge of $+\frac{2}{3}$ and the d quark $-\frac{1}{3}$.

3. *Spin "J."* Like electrons, quarks rotate around their own axes, which means they have their own angular momentum. The values of the momentum are multiples of a well-defined, smallest, nonzero angular momentum of an electron rotating around a proton in a hydrogen atom. The quarks rotate similarly to electrons and have an angular momentum of $\frac{1}{2}\hbar$. It is customary to set the constant equal to 1 and to measure all angular momenta in multiples of $\hbar$. Therefore, $+\frac{1}{2}\hbar$ becomes simply $+\frac{1}{2}$.

They can rotate in two opposite directions and this influences the weight of the nuclear particles formed such as protons, neutrons mesons, and many others.

Properties of Elementary Particles

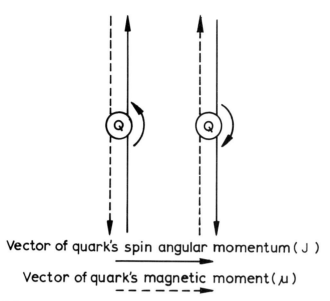

Vector of quark's spin angular momentum (J)

Vector of quark's magnetic moment (μ)

Figure 2.5. Solid arrow of quark's spin angular momentum (*J*); dashed arrow, vector of quark's magnetic moment (μ).

Mesons in which the spins of the quarks are parallel, are heavier than mesons in which the spins are opposed (Figure 2.5).

4. *Magnetic moment.* The spinning quarks also have a magnetic moment caused by the circulating electrical charge (or electric current). The vector of the magnetic moment is antiparallel to the spin vector (Figure 2.5).

2.4. BARYON NUMBER B

Baryons are composite particles (such as protons and neutrons) made up of three quarks, participate in the strong nuclear force interaction, and have a nonintegral spin of $\frac{1}{2}$, $\frac{3}{2}$, etc.

The major baryons are listed in Table 8.1 (see pp. 74–75). The baryon charge or number (*B*) is an important conservation law providing the stability of nuclear matter and of the proton. It says that the total number of baryons in a system, less the total number of antibaryons, must be conserved after interaction of particles. For example,

Proton	$B = 1$
Antiproton	$B = -1$
Mesons	$B = 0$

In a collision of two protons, for instance:

$$p + p \rightarrow p + n + p + \bar{\nu}$$
Baryon count $\quad 1 + 1 = 2 \quad 1 + 1 + 1 + -1 = 2$

The baryon charge of 2 has been maintained after the collision.

As protons and neutrons consist of three u, d quarks, the baryon charge B of u and d quarks is $\frac{1}{3}$.

Table 2.1 lists important characteristics of the stable elementary particles of which all matter in the universe is comprised.

The electron and neutrino are leptons—particles that do not participate in strong nuclear interactions. Quarks participate in strong interactions.

2.5. OTHER ELEMENTARY PARTICLES

In addition to the four stable elementary particles (see Table 2.1) that were created during the birth of the universe and are its building blocks, several unstable elementary particles are created in outer space by cosmic rays bombardment, or artificially in high-energy particle accelerators. Combined with some of the stable particles, they create heavy nuclear particles, which are also unstable and decay shortly after creation. The charged heavy leptons participate in electromagnetic and weak interactions. The neutral leptons (neutrinos) participate only in weak interactions.

2.5.1. Heavier Leptons

The *muon* (μ) is an unstable heavy elementary particle of negative charge, similar to the electron but 207 times heavier. Its rest mass is 0.106 GeV. The muon disintegrates within 1/1,000,000 of a second into an electron an electron-antineutrino, and a muon-neutrino (Figure 2.6).

The muon-neutrino and muon-antineutrino are denoted as ν_μ and $\bar{\nu}_\mu$. Their possible mass is larger than that of the electron-neutrino (probable mass 30 eV). The muon-neutrino may have a rest mass of 285 keV. A muon-neutrino in a reaction with a proton transmits a large portion of its energy and momentum to the proton, which becomes excited and disintegrates into several particles. However, the muon-neutrino reacts without changing its identity, called a *neutral current reaction* (Figure 2.7).

The *tau* (τ) is an unstable, heavy elementary particle of negative charge, similar to the muon but much heavier. Its rest mass is 1.78 GeV or 1.9 times heavier than a proton. Its lifetime is very brief, approximately 10^{-13} sec, and it

Properties of Elementary Particles

Table 2.1. Stable Elementary Particles—The Building Blocks of Matter

Name	Notation Particle	Notation Antiparticle	Mass GeV[a]	Mass Grams	Electric charge	Spin	Baryon number	Half-lifetime (sec)	Decay particle
Leptons									
Electron	e^-	e^+	0.00051	0.9×10^{-27}	∓ 1	$\pm \frac{1}{2}$	0	Stable	
Electron-neutrino	ν_e	$\bar{\nu}_e$	$\sim 30 \times 10^{-9}$[b]	0.535×10^{-35}	0	$+\frac{1}{2}$	0	Stable	
Quarks									
Up quark	u	$\bar{u}$	0.312	0.556×10^{-24}	$+\frac{2}{3}$	$\pm \frac{1}{2}$	$\frac{1}{3}$		Can decay into d
Down quark	d	$\bar{d}$	0.312	0.556×10^{-24}	$-\frac{1}{3}$	$\mp \frac{1}{2}$	$\frac{1}{3}$		Can decay into u

[a] 1 eV = 1.782×10^{-33} g, 1 GeV = 10^9 eV.
[b] Perhaps as large as 30 eV.

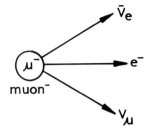

Figure 2.6. Disintegration of muon (μ) into an electron, electron-antineutrino, and muon-neutrino.

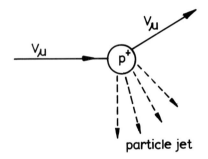

Figure 2.7. Neutral current reaction on muon-neutrino with a proton.

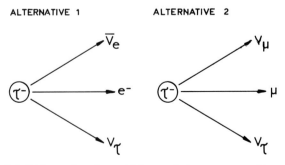

Figure 2.8. Disintegration of a tau (τ) into an electron or muon and two neutrino.

decays through the weak interaction similar to a muon into an electron, an electron-antineutrino, and a tau-neutrino. It can also decay into a muon, a muon-antineutrino, and a tau-neutrino (Figure 2.8). The tau-neutrino and tau-antineutrino, denoted ν_τ and $\bar{\nu}_\tau$, have a possible rest mass of 35 MeV.

2.6. HEAVIER QUARKS

In addition to the basic quarks u and d found in nucleons (protons and neutrons, which are the building blocks of the universe) together with electrons, neutrinos, and radiation, other types of much heavier quarks have been discovered.

2.6.1. Strange Quark, s

The discovery in 1950 of a heavy, short-lived, baryon named Λ hyperon, a particle with a rest mass of 1116 MeV or approximately 20% heavier than the proton, brought about the discovery of a new type of quark called strange (s) quark. The hyperon turned out to be a system consisting of three different quarks, u, d, and s. The s quark has an electrical charge of $-\frac{1}{3}$, which is the same as that of the d quark but with a rest mass of 0.45 GeV a sort of heavier cousin of the d quark. The short-lived particles containing the s quark are heavier than the particles containing only u and d quarks. There is of course an anti-s quark, $\bar{s}$.

2.6.2. Charm Quark, c

In 1974 at Brookhaven and SLAC simultaneously, a new meson called J/Ψ was discovered in proton–nucleon collisions. The particle was extremely heavy with a rest mass of 3097 MeV and a life span of 10^{-20} sec. It was later determined that the J/Ψ meson was a system of two new heavy quarks, which were given the name *charm* (c). The c quark has an electrical charge of $+\frac{2}{3}$ and seems to be a heavier cousin of the u quark. Its rest mass is 1.5 GeV or more than 50% heavier than a proton. The antiquark is denoted $\bar{c}$.

Subsequently, other heavier mesons were discovered in high-energy particle collisions where the charm quark combined with $\bar{u}$, $\bar{d}$, and $\bar{s}$ antiquarks. The particles are called D mesons, K particles, and F mesons. Theoretically, the c quark can bind with two other quarks such as u and d and form baryons. Such a particle was finally found in 1979 and denoted as Λ_c with a rest mass of 2273 MeV. It decays into $\Lambda \pi^+ \pi^+ \pi^-$ or $p K^- \pi^+ u$.

Table 2.2 Leptons and Quarks

Particle name	Notation Particle	Notation Antiparticle	Mass (GeV)	Q	B	L	L'	L"	J	T_Z	Half-lifetime (sec)	Decay particles
Leptons												
Electron	e^-	e^+	0.00051	± 1	0	± 1	0	0	$\tfrac{1}{2}\pm$	—	Stable	
Positron												
Muon	μ^-	μ^+	0.106	± 1	0	0	± 1	0	$\tfrac{1}{2}\pm$	—	2.2×10^{-6}	$e^- + \bar{\nu}_e + \nu_\mu$
Tau	τ^-	τ^+	1.78	± 1	0	0	0	± 1	$\tfrac{1}{2}\pm$	—	2.8×10^{-13}	$e^- + \bar{\nu}_e + \nu_\tau$; $\mu + \bar{\nu}_\mu + \nu_\tau$
Electron-neutrino	ν_e	$\bar{\nu}_e$	$\sim 0{-}30 \times 10^{-9}$	0	0	0	0	0	$\tfrac{1}{2}$	—	Stable	
Muon-neutrino	ν_μ	$\bar{\nu}_\mu$	~ 0.106	0	0	0	0	0	$\tfrac{1}{2}$	—	Stable	
Tau-neutrino	ν_τ	$\bar{\nu}_\tau$	~ 1.78	0	0	0	0	0	$\tfrac{1}{2}$	—	Stable	
Photon, quantum	γ		0	—					$1-$		Stable	
						s	c		T	T_Z		
Quarks												
Up quark	u	$\bar{u}$	0.312	$\pm \tfrac{2}{3}$	$\tfrac{1}{3}$	0	0		$\tfrac{1}{2}$	$\tfrac{1}{2}$		Can decay into a d quark
Down quark	d	$\bar{d}$	0.312	$\mp \tfrac{1}{3}$	$\tfrac{1}{3}$	0	0		$\tfrac{1}{2}$	$\tfrac{1}{2}$		Can decay into a u quark
Strange quark	s	$\bar{s}$	0.45	$\mp \tfrac{1}{3}$	$\tfrac{1}{3}$	-1	0		0	0		Can decay into a d or c quark
Charm quark	c	$\bar{c}$	1.5	$\pm \tfrac{2}{3}$	$\tfrac{1}{3}$	0	1		0	0		Can decay into an s, u, or d quark
Bottom quark	b	$\bar{b}$	4.9	$\mp \tfrac{1}{3}$	$\tfrac{1}{3}$	0	0					c quark + W^-; $\bar{u}d$, $c\bar{s}$, ν_e, e^-; $\bar{\nu}_\mu \mu^-$ or $\bar{\nu}_\tau \tau^-$
Top quark[a]	t	$\bar{t}$	>18	$\pm \tfrac{2}{3}$	0	0	0					

[a] This quark is predicted based on theoretical considerations but has not yet been discovered.

However, the c quark couples mainly with s, with a 5% chance of coupling with d.

2.6.3. Bottom Quark, b

In 1977–1978, a new, extremely heavy meson called upsilon Y, with a rest mass of 9460 MeV, was discovered in electron–positron annihilation. The particle behaved similarly to the J/Ψ meson and it was clear that it consisted of a new heavy quark denoted b (bottom) and its antiquark. It has a charge of $-\frac{1}{3}$, like the d and s quarks. The Y meson is a $b\bar{b}$ system. The rest mass of the b quark is 4.9 GeV or more than 5 times the weight of a proton.

2.6.4. Top Quark, t

In order to restore symmetry between leptons and quarks

$$\text{Leptons} \quad \begin{pmatrix} \nu_e \\ e^- \end{pmatrix} \begin{pmatrix} \nu_\mu \\ \mu^- \end{pmatrix} \begin{pmatrix} \nu_\tau \\ \tau \end{pmatrix}$$

$$\text{Quarks} \quad \begin{pmatrix} u \\ d \end{pmatrix} \begin{pmatrix} c \\ s \end{pmatrix} \begin{pmatrix} t \\ b \end{pmatrix}$$

another quark has been predicted. It would have to have a rest mass of 19–35 GeV and has been denoted t (top) quark. Sometimes the b and t quarks are called *beauty* and *truth*. So far, there is no concrete evidence that t exists. (See Table 2.2.)

Particles that have a spin $\frac{1}{2}$ are called fermions. The fermions appear in families. Only the first family (electron, electron-neutrino, and u and d quarks) appears in our universe as the building blocks of all matter. The second and third generations of heavier particles build up a new type of heavier matter, not found in nature, but produced in high-energy laboratories. This heavy matter decays rapidly into stable well-known particles of matter or radiation.

$$\begin{array}{ccc} \text{First} & \text{Second} & \text{Third} \\ \text{family} & \text{family} & \text{family} \\ \begin{pmatrix} \nu_e & u \\ e^- & d \end{pmatrix} & \begin{pmatrix} \nu_\mu & c \\ \mu^- & s \end{pmatrix} & \begin{pmatrix} \nu_\tau & t \\ \tau & b \end{pmatrix} \\ \text{stable} & \multicolumn{2}{c}{\text{unstable}} \end{array}$$

3

Basic Principles of Quantum Mechanics

In order to understand the properties, behavior, and interactions of elementary particles of matter, it is necessary to review at least the basic principles of the Planck theory of quantum mechanics.

At very small distances, smaller than 10^{-19} cm, nuclear particles do not behave in accordance with Newtonian mechanics. In other words, our physical concepts and principles of classical mechanics, based on experiences in the macroscopic world, are not suitable for analyzing something as tiny as an electron or quark.

To understand quantum mechanics is much more difficult than even the theory of relativity, which, in itself, is extremely complex. I do not expect the reader to fully understand quantum mechanics or present here the entire theory, but rather explain those parts that are relevant to the behavior of particles and the birth of the universe.

There are basically four important principles of quantum mechanics, proven experimentally and which apply to the behavior of nuclear particles at small distances: the quanta of electromagnetic energy, the uncertainty principle, the Pauli exclusion principle, and the wave theory of particles of matter.

3.1. THE QUANTA OF ELECTROMAGNETIC ENERGY

In 1900 Max Planck discovered that light or electromagnetic radiation is emitted in small but definite quanta of energy known as photons. In 1906 Einstein showed that light remains in these small definite quanta of energy when traveling through the universe.

Planck's constant reflecting the small quanta of energy is of fundamental importance to elementary particle physics. Its value is $h = 6.6 \times 10^{-34}$ W-sec.^{2}. It was first introduced by Planck in his theory of blackbody radiation. The watt (W) is the unit of power equal to 10^7 ergs/sec.

Planck's constant is mostly used in its reduced form $\hbar$:

$$\hbar = \frac{h}{2\pi} \tag{3.1}$$

or

$\hbar = 1.0546 \times 10^{-27}$ g-cm^2/sec $= 2.612 \times 10^{-66}$ cm^2 $= 6.6 \times 10^{-21}$ MeV-sec

It is extremely small, which explains why we can neglect quantum mechanics in the macroscopic world. Planck's constant appeared in Einstein's 1905 theory of photons, which specifies that the energy of a photon E_{ph} is Planck's constant $\hbar$ times the speed of light c divided by the wavelength λ:

$$E_{ph} = \frac{\hbar c}{\lambda} = \frac{\text{erg} \times \text{cm}}{\text{cm} \times \text{sec}} = \text{erg/sec} \tag{3.2}$$

The shorter the wavelength of the electromagnetic energy, the larger is the energy carried by the photon.

Later, Bohr discovered that the energies of electrons in atoms, as well as their momentum or spin, are also quantized and that both the permissible energies of electron orbits as well as the angular momentum or spin are controlled by the Planck constant $\hbar$.

Applying this theory to the simplest atom of hydrogen with one proton as nucleus and one orbiting electron, it was determined that if the electron stays in its standard orbit, no energy is radiated. However, when the electron is excited by heat or an electric current and jumps to a higher orbit, it radiates the required energy in the form of light shortly after descending to its basic orbit. When Bohr calculated the wavelength, he found that it corresponded to the spectrum of hydrogen (Figure 3.1).

The emission of electromagnetic energy (light) by the electron returning from the excited orbit to its permitted orbit can be detected as spectral light corresponding to the energy level of the electron orbit.

3.2. THE UNCERTAINTY PRINCIPLE

To determine at the same time the location and velocity of an elementary particle or atom presents a great problem. For example, if we take a moving electron and hit it with light or photons, the electron will reflect, first of all, the

Basic Principles of Quantum Mechanics

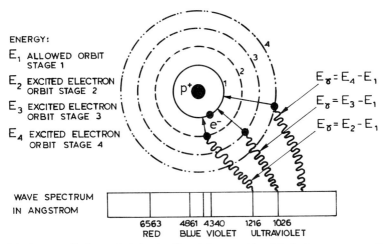

Figure 3.1. An atom of hydrogen depicting the allowed electron orbit and the hydrogen spectrum. Visible light spans from 3500 to 7000 angstroms (Å) (1 Å = 10^{-8} cm).

electromagnetic light waves, but, under the influence of the momentum and energy transferred by the beam of light waves, it will change its direction and velocity (Figure 3.2). For this reason, this theory, which is called the uncertainty principle of Heisenberg, determined that it is impossible to simultaneously establish the velocity and location of the electron in space with complete accuracy.

There is, however, a well-defined relationship between the various limits of uncertainties governed by the quantum theory, which claims that while the position and velocity of the particle cannot be established separately, the particle has a quantum state, which is a combination of position and velocity. If the position of a particle is known to be within a distance d from a point, then its momentum must be indeterminate by at least an amount i, and the relationship is governed by Planck's constant $\hbar$ as follows:

$$i \times d = \hbar$$

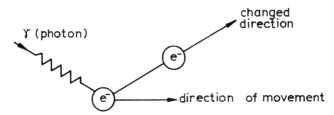

Figure 3.2. A moving electron hit by a photon.

The uncertainty principle can also be applied to Energy E and time t. Small amounts of energy ΔE can be transmitted for a very short time Δt, in violation of the energy conservation law.

$$\Delta E \times \Delta t = \hbar = 6.6 \times 10^{-21} \text{ MeV} \times \text{seconds} \qquad (3.3)$$

or

$$\Delta t = \frac{6.6 \times 10^{-21} \text{ MeV}}{\Delta E} \text{ seconds} \qquad (3.4)$$

If $\Delta t = 1$ sec, $\Delta E = 6.6 \times 10^{-21}$ MeV.

Based on the Heisenberg uncertainty principle, the larger the energy carried by an electron that released a virtual photon and violated the conservation law, the faster the balance must be restored. What happens here is that the photon will travel a shorter distance before being absorbed by another electron restoring the balance.

Considering that $E = mc^2$, time t will be

$$t = \frac{\hbar}{mc^2} \qquad (3.5)$$

Since particles of matter cannot travel at the speed of light, the absolute maximum distance would be ct:

$$ct = \frac{\hbar}{\text{mass (in MeV) } c} \qquad (3.6)$$

Another example of the uncertainty principle is the decay of free neutrons. We know that 50% (the half-life) of free neutrons will decay in approximately 15 minutes, but we cannot determine with certainty which particular neutron will decay.

3.3. THE PAULI EXCLUSION PRINCIPLE

The third principle of quantum mechanics was discovered by Pauli and states that two particles with spin $\frac{1}{2}$ such as electrons in atoms, quarks in protons, neutrons, and other baryons cannot occupy precisely the same quantum state. In other words, the Pauli theory, which is called the exclusion principle, forbids two electrons from occupying the same orbit around the nucleus in an atom.

It also applies to quarks and the way with which they can cluster together to form a proton or neutron. For example, in a proton the two u quarks spin in a parallel direction while the third d quark must spin in the opposite direction so the

Basic Principles of Quantum Mechanics

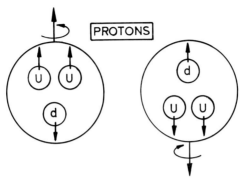

Figure 3.3. Pauli exclusion principle. The two u quarks must spin in an opposite direction to the d quark for the proton to have spin $\frac{1}{2}$.

proton can have a spin $\frac{1}{2}$ (Figure 3.3). But, as we will see later, the two u quarks must have a different charge, which is called color.

The exclusion principle also limits the electron in a hydrogen atom to move permanently on a specific and given orbit around the proton. If an electron switches to a higher orbit energized by heat or an electric current, it will return in less than 10^{-8} second to its basic orbit and will release the absorbed energy in the form of photons. A practical example of this phenomenon is the neon tube. Here, electrons in the atoms of the neon gas are excited by input of an electric current to move to larger orbits. Following the Pauli principle, they must, however, fall instantly into their stationary orbits, releasing electromagnetic radiation in the form of light.

3.4. THE WAVE THEORY OF ELEMENTARY AND SUBATOMIC PARTICLES IN GENERALIZED QUANTUM MECHANICS

Quantum mechanics describes the electrons circling a nucleus in an atom, not as pointlike particles but as superpositions of waves, which can be seen as a probability distribution around the nucleus. The energy of each distribution has a fixed value. Radiation at determined frequencies is emitted by the electron which jumps from one such quantum state to another.

Inspired by Planck and Einstein, de Broglie in 1925 expanded Planck's quantum electromagnetic wave theory to particles of matter. He introduced the notion that an electron moving on a given path exhibits wavelike characteristics, i.e., an electron wave (Figure 3.4).

Figure 3.4. The electron wave property in quantum mechanics.

Just as the relativistic theory of particles reflects the universal equivalence and interchangeability of mass and energy ($E = mc^2$), one of the fundamental properties of the generalized quantum theory is the inseparable connection between elementary particles and waves. In other words, quantum properties for particles are wave properties termed de Broglie waves.

The lowest energy orbit, called the Bohr orbit, fits exactly one wavelength. The wave peak falls into a trough and back to a peak, exactly at the point where the orbit of the electron started. When the de Broglie electron wave fits exactly the Bohr lowest energy orbit, the wave persists and ensures the proper orbiting of the electron. On the other hand, if the de Broglie wave does not fit exactly the orbit (higher energy orbit), the wave starts interfering with itself, dies out, and the electron returns to its Bohr orbit, emitting the extra energy in the form of photons. The higher the energy of the electron, the smaller the wavelength is.

The de Broglie wave theory explains the mechanics of why an electron, excited by heat or electric current energy and which moved to a higher energy orbit, falls instantaneously back to Bohr's allowed standard level. Only when one electron wave fits exactly the circle of the Bohr orbit does the electron remain steadily in the orbit. Higher energy orbits correspond to a large number of wavelengths that fit into the circumference of an orbit, interfere, die out, and the electron returns to its basic wave orbit.

Other subatomic particles such as neutrons, which in classical physics are considered to be pointlike particles and have mass, show characteristics of waves as well. A neutron acts in the same way as any other electromagnetic wave particle in the form of photons. When two waves of equal amplitude meet, they interimpose. When they are exactly in phase, they interfere and the resulting amplitude is twice as great. When they are exactly out of phase, they interface destructively and cancel out. The principle has been proven in the famous two-slit experiments with a neutron flux, as shown in Figure 3.5.

Each neutron shows at the same time its pointlike properties as it lands on the target screen. The type of the neutron wave determines the probability where a neutron will land. The probability is proportional to the square of the amplitude of the neutron's wave function. The wave properties of particles are evident on an atomic scale of 10^{-8} cm. A neutron wave, just as any other wave, has an amplitude and a phase that can be described as a wave function. The propagation

Basic Principles of Quantum Mechanics

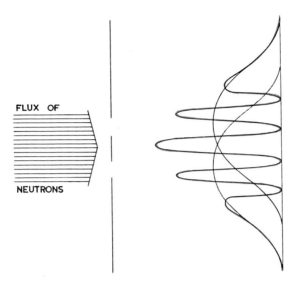

Figure 3.5. Two-slit experiment with neutrons. Based on the uncertainty principle, we cannot say that a neutron passed through one or the other slit. There are some spots on the screen where many neutrons land, other spots where no particle lands. The effects of the two slits can superimpose or cancel the waves.

of the particle waves follows the basic quantum-mechanical Schrödinger equation. The propagation of a neutron wave at room temperature is approximately 10^5 cm/sec.

The wavelength of a particle λ

$$\lambda = \frac{\hbar}{p} \qquad (3.7)$$

where $\hbar$ is Planck's constant = 6.625×10^{-27} erg·sec and p is the particle's momentum; and m_n is the mass of a neutron:

$$p = m_n \times v \qquad (3.8)$$

where v is the velocity

$$\lambda = \frac{\hbar}{m_n \times v} \qquad (3.9)$$

In gravitational experiments it has been proven that neutron waves are also influenced and bent by gravity. This is discussed in more detail in the chapter dealing with modern theories of gravity.

The main physical properties that characterize a wave are ω = frequency and

λ = length. To indicate the wavelength and its direction of propagation, it is necessary to introduce the magnitude of a wave vector:

$$V = \frac{2\pi}{\lambda}$$

The energy and momentum of a free-moving particle depend on its wave frequency ω and the wave number or magnitude:

$$E = \hbar\omega \tag{3.10}$$

$$p = \hbar V \tag{3.11}$$

or

$$E = \frac{p}{V}\omega \tag{3.12}$$

The frequency ω is measured in radians per second and is related to the frequency ν or oscillations per second:

$$\omega = 2\pi\nu \tag{3.13}$$

The equation $E = (p/V)\omega$ for energy and momentum now has a universal meaning. It established the relation between a given particle and its wave or the wave and particle.

This dual characteristic of particles is difficult to comprehend from the standpoint of classical physics. A particle is in a certain location or point, while waves propagate in space.

To understand this phenomenon, we must reconcile ourselves to the fact expressed in the Heisenberg uncertainty principle that we cannot establish with certainty in the microworld that a particle with a momentum p is located at a given point and so the quantum properties of a particle are waves and processes just as electromagnetic waves are particles.

From Eqs. (3.8) and (3.10) we can determine the relationship between frequency and wavelength. The kinetic energy for speeds not close to the speed of light is E_k; M_p is the mass of a proton and v its velocity:

$$E_k = \frac{M_p v^2}{2} \tag{3.14}$$

$$p = M_p v, \qquad p^2 = M_p^2 v^2, \qquad v^2 = \frac{p^2}{M_p^2}$$

$$E_k = \frac{p^2}{2M_p}, \qquad \lambdabar = \frac{\lambda}{2\pi} \tag{3.15}$$

$(E = \hbar\omega, p = \lambdabar v)$

Basic Principles of Quantum Mechanics

$$\lambda\omega = \frac{\hbar v}{2M_p}$$

$$\omega = \frac{\lambda v^2}{2M_p} = \frac{2\pi^2\lambda}{\lambda^2 M} \tag{3.16}$$

$$\lambda = \frac{\pi \lambda 2^{1/2}}{(M_p E)^{1/2}} \tag{3.17}$$

The wavelength of a particle is inversely proportioned to the square root of energy.

For the light quantum or photon moving at the speed of light, its energy E, momentum p, and wavelength λ are:

$$E = cp \tag{3.18}$$

$$p = \frac{2\pi c}{\lambda} \tag{3.19}$$

$$E = \frac{2\pi c\hbar}{\lambda}$$

$$\lambda = \frac{2\pi c\hbar}{E} \tag{3.20}$$

We can therefore determine that for different particles with equal wavelengths we obtain different energies.

Lighter particles have more pronounced wave properties. It is more convenient in quantum mechanics to use a notation $\lambdabar$ for wavelength:

$$\lambdabar = \frac{\lambda}{2\pi}$$

The wave vector V becomes

$$V = \frac{1}{\lambdabar}$$

and energy E with M the mass of the particle:

$$E = \frac{\lambdabar^2}{2M\lambda} \tag{3.21}$$

The question now is, what are the limitations of classical mechanics and when do quantum mechanics apply?

According to the Heisenberg uncertainty principle, the limits of applicability of classical concepts are determined by

$$\Delta X \times \Delta p \geq \frac{\lambda}{2\pi} \tag{3.22}$$

where ΔX is the uncertainty in the value of coordinates and Δp is the uncertainty in the value of momentum.

A similar relationship exists between energy E and time t:

$$\Delta E \times \Delta t \geq \frac{\hbar}{2} \tag{3.23}$$

The meaning of all this is that if we measure at the same time the coordinate and the momentum of a particle, the errors will satisfy the uncertainty relation (3.22).

In classical laws of mechanics, the uncertainties expressed in Eqs. (3.21) and (3.22) must be satisfied.

We will now check whether, in accordance with this theory, electrons and protons are really quantum objects. Assuming that the coordinate and momentum of the particle oscillate at a zero average value, then $\Delta p = p$ and $\Delta X = X$, p and X denote the root mean squares of Xp and

$$2X^2 \times p^2 \geq \frac{\hbar^2}{4}$$

$$E = \frac{p^2}{2M} \quad \text{or} \quad p = (E2M)^{1/2}$$

$$X^2 \times p^2 \geq \frac{\hbar^2}{4} \quad 4EMX^2 \geq \frac{\hbar^2}{4}$$

$$X = \left(\frac{\hbar^2}{EM}\right)^{1/2} \tag{3.24}$$

For an electron in an atom $E \sim 10$ eV

$$\left(\frac{\hbar^2}{EM}\right)^{1/2} = 2 \times 10^{-8} \text{ cm}$$

For a proton or neutron $E \sim 10$ MeV

$$\left(\frac{\hbar^2}{EM}\right)^{1/2} = 4 \times 10^{-13} \text{ cm}$$

The condition of $X \geq (\hbar^2/EM)^{1/2}$ does not hold for electrons, protons, and neutrons; therefore, they are quantum objects. Also, the smaller the distance between particles, the higher the energies are. The investigations at short distances are investigations at high energies.

Other Planck constants are Planck energy E_p or Planck mass M_p, given by Newton's gravitational constant G and the Planck constant $\hbar$:

$$E_p = \left(\frac{\hbar c}{G}\right)^{1/2} = 1.1 \times 10^{19} \text{ GeV} = 2 \times 10^{-5} \text{ g in mass units} \tag{3.25}$$

It corresponds to the characteristic Planck length or distance:

$$L_p = \left(\frac{\hbar G}{c^3}\right)^{1/2} = 1.61 \times 10^{-33} \text{ cm} \tag{3.26}$$

and Planck density

$$\frac{M_p}{L_p^3} = \frac{c^5}{\hbar G^2} = 5.15 \times 10^{93} \text{ g/cm}^3 \tag{3.27}$$

3.5. APPLICATION OF PLANCK'S QUANTUM THEORY TO ELECTROMAGNETIC RADIATION

For each wavelength of electromagnetic radiation from a blackbody in thermal equilibrium such as a star, there is a corresponding minimum amount or quantum of radiation that cannot be split into smaller amounts. In addition, there is a firm relationship between the temperature of a radiating blackbody and the total energy radiated. It is directly proportionate to the fourth power, $T4$. It is clear from Figure 3.6 that if the temperature is 10 times higher, the total electromagnetic energy radiated will be 10^4 or 10,000 times larger.

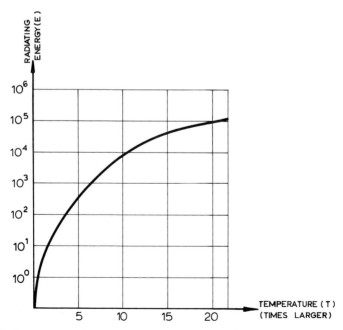

Figure 3.6. Star temperature in relation to radiation energy. $E = aT^4$ and $T = 10$ times larger, $E = a10^4 = 10,000$ larger.

The typical wavelength of blackbody radiation in equilibrium is 0.29 cm at a temperature of 1 K. The energy of radiation is proportionate to the temperature and inversely proportionate to the wavelength. From these two rules we can say that the wavelength of photons is indirectly proportionate to the temperature. Therefore, if

$$\lambda_0 = 0.29 \text{ cm for } 1 \text{ K}$$

for a temperature T, the wavelength will be

$$\lambda_T = \lambda_0/T = 0.29/T$$

This explains why we can see the surface of the sun with a temperature of about 5800 K. The radiation will appear on a wavelength λ_{SSR}:

$$\lambda_{SSR} = 0.29/5800 = 0.00005 = 5 \times 10^{-5} \text{ cm}$$

As 1 angstrom = 10^{-8} cm

$$\lambda_{SSR} \sim 5000 \text{ angstroms}$$

This wavelength represents the core of visible light.

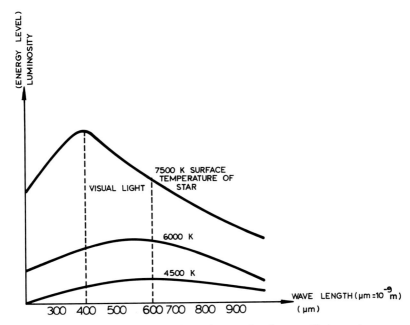

Figure 3.7. Star's energy radiated at various wavelengths at specific temperatures.

Basic Principles of Quantum Mechanics 33

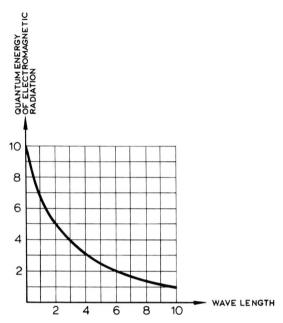

Figure 3.8. Wavelength received from a radiating star or galaxy versus quantum energy.

3.6. LUMINOSITY OF RADIATING BODIES SUCH AS STARS OR GALAXIES

The form of the luminosity curve depends entirely on the surface temperature of the light source. The higher the surface temperature of a star, the higher is the peak of the luminosity curve (higher energy).

Regardless of whether the radiation is from a close or far source, small or large objects, the radiation comes in quantum amounts and this depends only on the temperature of the source. The quantum energy depends on the wavelength and increases in direct relation to the reduction of wavelength (Figure 3.7). If the wavelength registered from a cosmic object (e.g., star, galaxy) is increased, the quantum energy of the radiation is reduced and vice versa, as shown in Figure 3.8.

4

Einstein's Relativistic Properties of Particles

At higher velocities and energy levels, the classical Newtonian mechanics do not apply to the behavior of particles. The relativistic mechanics or properties of particles based on Einstein's theory of relativity have little effect if the particle velocity v is small in relation to the speed of light $c \sim 3 \times 10^{10}$ cm/sec:

$$\text{For } v_{nonrel} \leq c, \quad E \leq Mc^2$$

The most important relativistic equation is the energy and mass equivalence, the basic equation of Einstein:

$$E_{rel} = Mc^2 \tag{4.1}$$

where M is the relativistic mass.

Another important relation is the connection between energy, mass, and momentum p:

$$E_{rel} = c(p^2 + M^2c^2)^{1/2} \tag{4.2}$$

where M is the rest mass.

For a particle at rest $p = 0$, $E_{rel} = Mc^2$ (this is also the rest mass of a particle). The relativistic kinetic energy of a particle E_k is the total energy less the rest energy:

$$E_k = c(p^2 + M^2c^2)^{1/2} - Mc^2 \tag{4.3}$$

$$\text{For } p \leq Mc, \quad E \cong \frac{p^2}{2M} \quad \text{(nonrelativistic)} \tag{4.4}$$

$$\text{For } p \geq Mc, \quad E \cong c(p) \quad \text{(relativistic)} \tag{4.5}$$

$$p \cong \frac{E}{c}$$

For example, a proton with a momentum of 100 MeV/c has a kinetic energy of 20 MeV and with a momentum of 1 GeV/c, has a kinetic energy of 400 MeV.

The velocity v of a particle moving at close to the speed of light is

$$v = \frac{p^2 c^2}{E_{rel}} \tag{4.6}$$

Substituting E_{rel} from Eq. (4.2)

$$v = \frac{pc^2}{c(p^2 + M^2 c^2)^{1/2}} \tag{4.7}$$

A photon or quantum of electromagnetic energy has a rest mass of $M = 0$. Equation (4.7) becomes

$$v_{ph} = \frac{pc^2}{cp} = c \tag{4.8}$$

This indicates that the only velocity a zero-mass particle such as a photon can have is the speed of light ($c = 3 \times 10^{10}$ cm/sec). No physical object having a mass $M > 0$ can move with the velocity of light. This does not, however, limit, theoretically, electromagnetic radiation of very high frequency to move faster than the speed of light.

Equation (4.2) can also be written

$$E^2_{rel} = c^2 p^2 + M^2 c^4 \tag{4.9}$$

$$E^2_{rel} - c^2 p^2 = M^2 c^4 \tag{4.10}$$

This equation can be explained as follows.

If energy and momentum of a particle are measured in one reference frame and subsequently in another reference frame that moves at a definite velocity in comparison to the original frame of reference, the individual values of E and p will be different; the value $E^2_{rel} - c^2 p^2$ will, however, remain the same in both frames of reference—*it is invariant*.

The invariant for two sets of (pEp) and (qEq) of momenta and energies will be the quantity $EpEq - c^2 pq$.

The Galileo law of summation of velocities in nonrelativistic physics is valid:

$$v_2 = v_1 + v \quad \text{(nonrel)}$$

where v_2, v_1 are measured in two reference frames where one moves at velocity v in relation to the other.

In relativistic physics, Einstein's law of summation is valid:

$$v_2 = \frac{v_1 + v}{1 + v_1 v_2 / c^2} \quad \text{(rel)} \tag{4.11}$$

The definition of rapidity y is introduced in relativistic kinematics:

$$y = \tfrac{1}{2}\ln\frac{c + v}{c - v} \qquad (4.12)$$

As can be seen, rapidity is determined only by velocity.

INFLUENCE OF CLOSE-TO-SPEED-OF-LIGHT VELOCITIES ON MASS, LENGTH, AND TIME

Increase of Mass

Based on Einstein's special theory of relativity, the mass of an object moving at speeds approaching the speed of light increases considerably and doubles at a speed of 250,000 km/sec or 83% of the speed of light (Figure 4.1).

A proton at rest has an energy of 1 GeV and weighs 10^{-24} g. Accelerated to close to the speed of light in a supercollider at 20 TeV, the weight of the proton increases to 3×10^{-20} g.

As a consequence of the theory of relativity and the equivalence of mass and energy ($E = mc^2$), the energy that an object has, due to its motion, adds to its mass. At 90% of the speed of light, the mass is more than twice the normal mass.

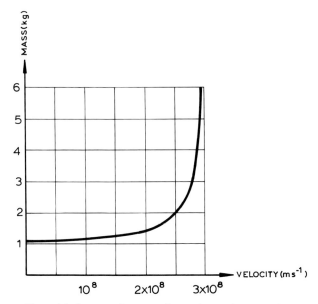

Figure 4.1. Increase of mass moving at close to the speed of light.

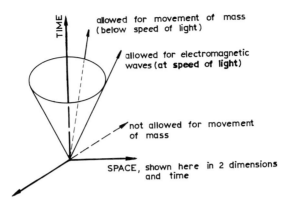

Figure 4.2. The light cone. The path of any object must move within the light cone in four-dimensional space time.

It takes more and more energy to increase the velocity and it would take an infinite amount of energy to bring the object to the velocity of light, c. The mass would then become infinitely large. Therefore, only electromagnetic energy in the form of photons (which have no mass) can move at the speed of light.

This can be expressed in graphical form by a so-called light cone in four-dimensional Einstein space-time (Figure 4.2). Massive particles can only move within the cone at speeds below the speed of light.

Slowdown of Time

In a similar relation to mass versus speed at close to the speed of light, time slows down when a body moves at close to the speed of light (Figure 4.3).

Einstein introduced in his theory of relativity the four-dimensional space-time—three coordinate dimensions, length, width and height (x, y, and z), and time (t). Prior to Einstein, Newton's view prevailed, where space and time existed separately without reference to anything external. Space consists of an infinite number of points and an event occurs in a specific point of the three-dimensional space at a specific time. Einstein combined the four in one inseparable continuum, space-time.

Time as a mathematical dimension was introduced into Einstein's theory by Minkowski as an imaginary complex number based on the square root of -1 or $\sqrt{-1} = i$. Time t, therefore, in Einstein's theory of relativity is written as

$$t_{\text{rel}} = ict \qquad (4.13)$$

Einstein's Relativistic Properties of Particles

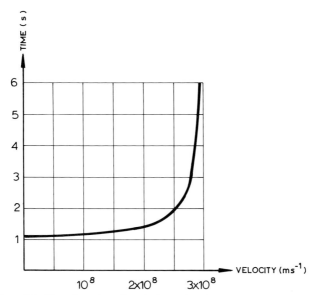

Figure 4.3. Time slows down for a body moving close to the speed of light.

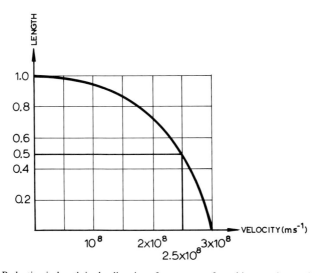

Figure 4.4. Reduction in length in the direction of movement of an object moving at close to the speed of light.

However, although time is entered as an imaginary or complex number, the results in measurement of time become real numbers. More will be said about this in the chapter describing twistor theory.

REDUCTION IN LENGTH

The length of an object decreases considerably in the direction of movement with speed. At 250,000 km/sec or 83% of c, the length of an object is shortened to 50% of its original value (Figure 4.4).

5

Quantum Electrodynamics

The quantum theory of electromagnetic interactions is called quantum electrodynamics (QED). The modern concept of the electromagnetic force between two moving electrically charged particles describes the interactions as taking place in two stages. First, a photon or quantum of electromagnetic radiation is emitted by one electron, shown in Figure 5.1, which is then absorbed by the other electron. The photon is the carrier or boson of the electromagnetic force. Another description of the same interaction is that the moving electron creates an electric current, similar to that in wire, which repels the other electric current created by the second electron. The electrical charge in both descriptions is preserved after the interaction.

The electrical charge and strength of the electromagnetic charge in particles, such as electrons or quarks, is a mystery that has not been fully understood. It is a property that is maintained during the interactions and does not change.

The electrical charge of an electron, positron, proton, and antiproton has the same value but opposite sign. An electron and antiproton have a negative charge -1; the proton and positron charge is $+1$. The positron can be accelerated to close to the speed of light in a cyclotron and its weight increased to that of a proton, but the charge remains the same, $+1$.

The electromagnetic attraction between negatively charged electrons and positively charged protons in the nucleus keeps the electrons in their orbit around the nucleus of atoms. Again, the electromagnetic force is caused by the exchange of virtual photons. However, when an energized electron moves to a higher orbit, it returns instantaneously to its permanent stationary orbit and releases a real photon in the form of light. A virtual or to be photon appears and disappears rapidly, usually in 10^{-23} sec.

The electrical charge of an electron or proton is $e = 1.6 \times 10^{-19}$ C (coulomb)

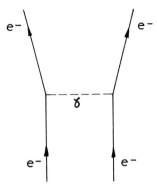

Figure 5.1. Interaction of two electrons. A γ photon is emitted by one electron and is absorbed by the other electron. The photon is virtual or to be.

and the strength of the electromagnetic interaction or the "fine structure constant" α of the quantum electrodynamics (QED) is

$$\alpha = \frac{e^2}{\hbar c} = \frac{1}{137} = 0.0072992 = 7.29 \times 10^{-3} \quad (5.1)$$

where

$$\hbar = h/2\pi = 6.6 \times 10^{-22} \text{ MeV-sec (Planck's constant)}$$

and

$$c = 3 \times 10^{10} \text{ cm/sec (speed of light)}.$$

The fine structure constant is the strength by which electrons couple to electromagnetic radiation. $\alpha = 7.29 \times 10^{-3}$.

If m_p and m_e are proton and electron masses in a hydrogen atom (Figure 5.2), then the ratio between gravitational and electromagnetic forces is

$$\frac{Gm_p m_e}{e^2} \cong 10^{-40} \quad (5.2)$$

Figure 5.2. Hydrogen atom.

Quantum Electrodynamics

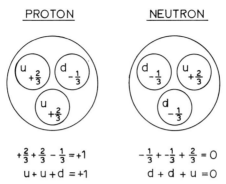

Figure 5.3. The electrical charge of protons and neutrons is the result of the electrical charge of its constituents, the quarks u ($+\frac{2}{3}$) and d ($-\frac{1}{3}$).

The electromagnetic force keeping the electron in its orbit around the proton and in general keeping atoms in equilibrium is 10^{40} times larger than gravity, which has little influence on particle interactions.

The electrical charges of quarks u and d, all of which are confined in protons and neutrons, are u = $+\frac{2}{3}$ and d = $-\frac{1}{3}$ with the resulting charge of +1 for a proton and 0 for a neutron (Figure 5.3). The repulsive and attracting electromagnetic forces in the proton and neutron confinement, however, play an insignificant role against the color–gluon strong nuclear force, which is so large that two quarks cannot separate beyond the radius of a proton. The strong nuclear force at a distance of 10^{-13} cm becomes so strong that it overcomes the repulsive electromagnetic force of protons, and thus the nucleus is formed at the core of an atom.

The main area of action of electromagnetic forces is the range of 10^{-12} cm to a few centimeters, including atomic, molecular, crystal structures, chemical reactions (exchange of electrons), thermal and physical properties of material and friction. The range extends to galaxies billions of light-years away (see Table 5.1).

The electromagnetic force interacts also to the same extent with particles of nonzero mass and spin via their magnetic moments, which are $e\hbar/2mc$, where m

Table 5.1. Intensity and Range of Interactions

Interaction force	Relative strength	Range
Strong	1	10^{-13} cm
Electromagnetic	10^{-3}	∞
Weak	10^{-24}	10^{-11} cm
Gravitational	10^{-40}	∞

is the mass of the particle (e.g., neutron). The neutrino and all neutral particles are immune to electromagnetic interactions.

Coulomb's Law

The force F_{ab} between two stationary electric point particles Q_a and Q_b acts along a line joining the two charges (Figure 5.4); it is proportional to the product $Q_a \times Q_b$ and inversely proportional to the square of the distance, d, separating the charges:

$$F_{ab} = \frac{1}{4} \frac{Q_a Q_b}{\epsilon_0 d^2} d_1 \qquad (5.3)$$

where d_1 is a unit vector pointing in the direction of Q_a and Q_b. The force is attractive if the charges have different signs $(-,+)$ and repulsive if the charges are of the same sign $(-,-)$. F is measured in newtons, Q in coulombs, and d in meters. ϵ_0 is the permittivity of free space:

$$\epsilon_0 = 8.854 \times 10^{-12} \text{ farad/meter}$$

Substituting with ϵ_0 in (5.3)

$$F_{ab} = 9 \times 10^9 \frac{Q_a Q_b}{d^2} d_1 \qquad (5.4)$$

The Coulomb forces are enormous in comparison to gravity.

The gravitational force between masses m_a and m_b at distance d is

$$G_F = \frac{6.672 \times 10^{-11} m_a m_b}{d^2} \qquad (5.5)$$

In comparison, the gravitational force of the sun on a proton on the surface (mass of the sun = 2×10^{30} kg, radius = 7×10^8 m) is equal to the electric force between a proton and an electron separated by the sun's radius of 7×10^8 m.

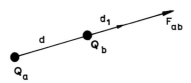

Figure 5.4. The force F_{ab} between two stationary pointlike particles a, b with electrical charge Q_a and Q_b.

Quantum Electrodynamics

The Electric Field Intensity E

If an electrically charged point particle such as an electron or proton Q_a stands motionless, it sets up an electric field extending in three dimensions (Figure 5.5). We define the electric field intensity E as a force per unit charge on a test charge in the field. The electric field intensity due to the point charge Q_a is

$$E_a = \frac{F_{ab}}{Q_b} = \frac{9 \times 10^9 Q_a}{d^2} d_1 \qquad \text{volts/meter} \qquad (5.6)$$

The electric field intensity due to point charge Q_a is the same whether the test charge Q_b is in the field or not and regardless of how large Q_b is.

Two Oppositely Charged Particles at Rest

A negatively charged and a positively charged particle such as an electron and positron pair, a quark and antiquark in a meson, attract each other by lines of force that fill the space between the particles (Figure 5.6).

Electrically Charged Particle in Motion

An electrically charged particle in motion sets up both an electrical and a magnetic field, differently oriented (Figure 5.7).

An Electron or Proton in Accelerated Motion

An electron or proton in accelerated motion creates an electromagnetic wave, which consists of oscillating electric and magnetic fields of the same intensity. In Figure 5.8, the electric field is vertical and the magnetic horizontal.

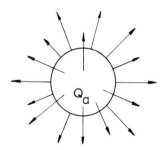

Figure 5.5. The electric field of charged particle Q_a extends into space in three dimensions.

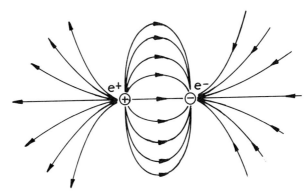

Figure 5.6. The electric field lines between two particles, e⁻ (electron) and e⁺ (positron), fill all space.

The merger of electricity and magnetism into a single theory was the work of James Clerk Maxwell in the 19th century. The Maxwell equations describe the propagation in space of electromagnetic quanta (quantum mechanics terminology) or photons. They describe all electromagnetic phenomena from galactic fields to minute interactions at 10^{-16} cm.

The Maxwell equations of electromagnetism expressed in vector calculus are

$$\overline{V}E = 4\pi p \tag{5.7}$$

where E is the electric field intensity and p is the density of the electrical charge. The current of the moving electrical charge is

$$-\frac{1}{c}\frac{\partial E}{\partial t} + \overline{V} \times M = \frac{4\pi}{c}J \tag{5.8}$$

where c is the velocity of light, J is the current flow of the electrical charge, ∂ are the partial derivatives of the electric field and time, and M is the magnetic field.

This also applies to all interactions of charged particles such as scattering of

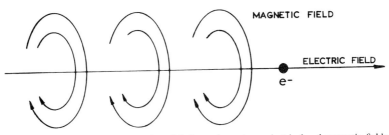

Figure 5.7. An electrically charged particle in motion sets up electrical and magnetic fields.

Quantum Electrodynamics

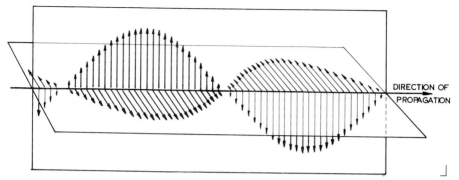

Figure 5.8. Propagation of an electromagnetic wave in space. Electric and magnetic fields are oscillating in equal intensity and perfect balance. Such a wave can be created by accelerating an electrically charged particle such as a proton or electron.

two electrons flying through space. The two electrons exchange photon quanta, in this case a virtual photon. This exchange is the cause of the electromagnetic repulsion of the two electrons (Figure 5.9).

Magnetic Monopoles

It is interesting to note that Maxwell's merger of electricity and magnetism and the fact that magnetism exists only as a consequence of the motion of electrically charged particles, creates in actuality an asymmetry of electricity and magnetism. There are those who claim that for the sake of symmetry, there should exist in nature magnetic particles that create a magnetic field and in motion produce electric fields in the same way as is done by moving electrically

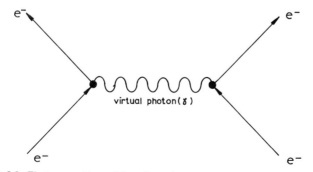

Figure 5.9. Electromagnetic repulsion of two electrons by exchange of a virtual photon.

charged particles. So far, the extensive search for these magnetic particles, called monopoles, has been unsuccessful.

Encounter of an Electron and a Positron Called Positronium: Direct Transfer of Mass into Energy

When an electron (charge, −1) encounters its antiparticle positron, which has the same mass but an opposite electrical charge (+1), the system is called a positronium. As all particles and antiparticles annihilate each other, a positronium or a bound state of e^+ and e^- decays into electromagnetic radiation:

$$e^- + e^+ \to \gamma \text{ (photons)}$$

The transit time t_{ep}, the time it takes for the particles to penetrate the other particles, is calculated from the equation

$$t_{ep} = \frac{R_e}{v} \tag{5.9}$$

where R_e is the radius of an electron or 10^{-17} cm. At a particle velocity of $v = 0.99c$ or $0.99 \times 3 \times 10^{10}$ cm,

$$t_{ep} = \frac{10^{-17}}{2.97 \times 10^{10}} = 3.3 \times 10^{-25} \text{ sec}$$

Depending on the energies of the particles, the speed may be lower; then the penetration time is shorter.

The decay that produces pure electromagnetic energy (γ) is the most remarkable illustration of Einstein's theory of direct conversion of mass into energy, $E = mc^2$.

Depending on the spin of the two particles, we obtain two types of positronium: *parapositronium*, the particles spinning in opposite directions (Figure 5.10), and *orthopositronium*, the particles spinning in the same direction (Figure 5.11). Parapositronium has an orbital angular momentum of 0. Orthopositronium has a total angular momentum of $2 \times \frac{1}{2} = 1$. Two photons cannot produce an angular momentum of 1 in units of $\hbar$ (Planck's constant). The process must therefore produce three or more photons. The decay rate of parapositronium is $\alpha^2 = 1/137$ (two photons) while in orthopositronium it is $\alpha^3 = 1/137$. The decay rate is therefore reduced by $\alpha = 1/137$ and for this reason the lifetime of orthopositronium is 100 times longer, 10^{-8} versus 10^{-10} sec, than that of parapositronium.

The theoretical cross section of annihilation σ_{ann}, i.e., the area being struck in which the two flying particles annihilate, can be calculated from the following equation:

Quantum Electrodynamics

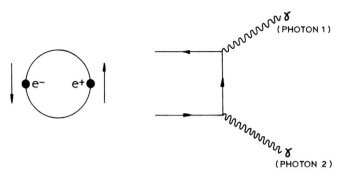

Figure 5.10. *Parapositronium*. The electron (e⁻) and positron (e⁺) spin in opposite directions. Their annihilation produces two photons.

$$\sigma_{ann} = \pi r_e^2 \frac{c}{v} \tag{5.10}$$

where v is the velocity of the positron and r_e is the radius of an electron. At $v = 0.9c$

$$\sigma_{ann} = 3.14 \times 10^{-26} \times 0.9 = 2.8 \times 10^{-26} \text{ cm}^2$$

The possible interactions of electrons and positrons are the following:

$$\gamma + e^- \rightarrow \gamma' + e^{-\prime}$$
$$e^- + e^- \rightarrow e^{-\prime} + e^{-\prime}$$
$$e^- + e^+ \rightarrow e^{-\prime} + e^{+\prime}$$
$$e^- + e^+ \rightarrow \gamma' + \gamma'$$

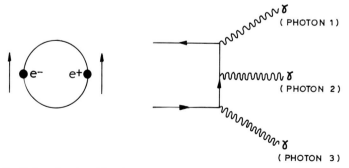

Figure 5.11. *Orthopositronium*. The electron (e⁻) and positron (e⁺) spin in the same direction. Their annihilation produces three photons.

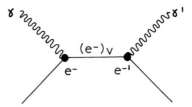

Figure 5.12. Scattering of photons by electrons. The virtual electron $(e^-)_v$ cannot exist and decays into electronmagnetic energy γ' and a real electron moving in a different direction.

$$\gamma + e^+ \rightarrow \gamma' + e^{+'}$$
$$e^+ + e^+ \rightarrow e^{+'} + e^{+'}$$
$$\gamma + \gamma \rightarrow e^- + e^+$$

Figure 5.12 shows the principal interaction of scattering of photons by electrons. The virtual electron (e^-) cannot exist and decays into γ' quantum of electromagnetic energy and a real electron, which will move in a different direction. $\gamma + e^- \rightarrow (e^-)_v \rightarrow \gamma' + e^{-'}$. This is called a secondary reaction via a virtual process. The photon transfers a momentum to the electron and changes its direction. The photon as well changes its wavelength and energy and scatters in a different direction.

Considering that both the electron and the photon in quantum mechanics are waves (Compton effect), the scattering as described has two effects on the electron and photon waves: (1) it increases the Compton wavelength of the electron by transfer of part of the photon's energy and (2) it decreases the frequency of the photon wave due to the decrease of energy transferred to the electron.

The Compton wavelength of the electron,

$$\Lambda = \frac{\hbar}{mc} \sim 4 \times 10^{-11} \text{ cm}$$

serves as the scale of length for quantum electrodynamics.

6

Quantum Chromodynamics, the Strong Nuclear Force

6.1. INTERACTIONS OF QUARKS VIA THE STRONG NUCLEAR FORCE

There are no free quarks in the universe. All quarks—free in the fireball during the birth of the universe—are now confined to protons and neutrons called baryons. The combination of three of the u and d quarks, held together by the strong nuclear force and their carrier (the gluons), formed the protons and neutrons, the nucleus of atoms. In cosmic interactions and mainly in high-energy particle accelerators, other combinations of nuclear particles called baryons and mesons are created, but they are all extremely short-lived and decay in protons, neutrons, electrons, neutrinos, and photons. Mesons are combinations of one quark (q) and one antiquark ($\bar{q}$), again held together by the strong force, but are short-lived.

Free neutrons, not confined to atoms, decay in approximately 15 minutes. The present generally recognized theory of quark interactions is called quantum chromodynamics (QCD) and we will explain this theory in more detail.

To account for the confinement of quarks in nuclear particles such as protons (two u + one d quark), neutrons (two d + one u quark), and mesons (one quark + one antiquark), a dynamic property of quarks has been established, which is called *color*. Each type or flavor of the two quarks u and d comes in three colors: red, green, and blue. Antiquarks have anticolor: antired, antigreen, and antiblue. There are other types of quarks or flavors such as s, c, b, and t but those heavy quarks appear only in short-lived baryons and mesons, and are not combined in the matter that fills the universe. The quark color property has nothing to do with visual color. The term is used because the way the different colored quarks combine in confinement in baryons is similar to the way visual colors combine. In other words, the three quarks in a proton, which is "white" or colorless, each have a different color—red, green, and blue—which, when combined, give a neutral colorless property to the proton.

Just as u and d quarks have a spin angular momentum of $\pm\frac{1}{2}$ that can be oriented up or down in reference to any spatial axis, quarks have different discrete properties and values.

All nuclear particles, which consist of quarks and antiquarks and participate in the strong interaction, are called hadrons. Hadrons are divided into baryons such as the proton and neutron, which consist of three quarks, have nonintegral spin ($\frac{1}{2}$, $\frac{3}{2}$, etc.), and obey the Pauli exclusion principle, and mesons, which consist of a quark and an antiquark, have integral spin (0, 1, 2, etc.), and do not obey this principle. The Pauli exclusion principle, which is physically obeyed also by leptons (electron, muon, tau, and neutrino), states that two elementary particles (quarks, leptons) of the same type such as two u quarks in a proton with nonintegral spin cannot occupy the same quantum state, as mentioned before.

All hadrons are white (colorless), consisting of three quarks in three colors—red, green, and blue, which cancel out. Mesons, which consist of a colored, say red, quark and an antiquark, in this case antired, are also colorless. The strong nuclear force responsible for confining the quarks in hadrons based on the quark properties of color is in this way similar to electromagnetic interaction, which is based on the electrical charge. There is no color preference in nature. All three color quarks are equally numerous.

The quark color theory was first established to explain a heavy, short-lived particle, Δ^{++}. This unstable particle created in high-energy accelerators consists of three u quarks—all with the same direction of spin—and has an electrical charge of $+2$ ($+\frac{2}{3} + \frac{2}{3} + \frac{2}{3}$) and a mass of 1.2 GeV. In order to comply with the Pauli exclusion principle, the u quarks (spin $\frac{2}{3}$), which all spin in the same direction, should be in an antisymmetric configuration but in the Δ^{++} particle they are in a symmetric position, all spinning in the same direction (Figure 6.1).

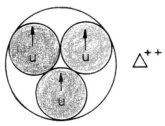

Figure 6.1. Δ^{++} particle has three u quarks spinning in the same direction, seemingly violating the Pauli principle. The u quarks have three different characteristics and the Pauli principle is maintained.

Quantum Chromodynamics

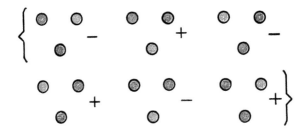

Figure 6.2. The SU(3) chromodynamic (QCD) group. Shown are all possible combinations of color characteristic or charge for the three quarks of a proton, neutron, or other baryon.

The explanation for this phenomenon is that the u quarks come in three different color combinations and thus have three different charges or characteristics and, therefore, the Pauli exclusion principle can be maintained (Figure 6.2).

The Δ^{++} configuration as shown is in a superposition of six different sets with alternating signs. The sum of all sets is antisymmetric when two u quarks of different colors are interchanged. All quark configurations are antisymmetric with respect to color and each color enjoys the same rights as the others. In mathematical terms, these are *color singlets*. As we have three colors, the group is called SU(3), describing all possible arrangements of three colors, as shown in Figure 6.2.

Just as baryons are called color singlet configurations, the same applies to mesons, which are combinations of a quark and antiquark (Figure 6.3). A meson is the sum of three configurations red + green + blue and the sum of all three sets is again a color singlet which is called colorless or "white."

In nature, only color or charge singlets exist as real particles. The color characteristic of quarks can be interpreted as being similar to electrical charge.

As explained in electrodynamics (QED), particles of opposite charge, such as an electron (−) and proton (+), attract each other, whereas two electrons having the same charge repel each other (Figure 6.4). An electron and positron (electron with positive charge) attract each other and form what is called a bound system with charge 0 [(+) + (−) = 0] or positronium. This may be interpreted as

Figure 6.3. A meson consists of a quark and antiquark and is the sum of three configurations of colors or charges.

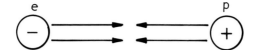

Figure 6.4. Two particles of opposite charge attract each other.

a charge singlet in a similar way as a color singlet with quarks, and the resulting charge is $+2$ or -2. There is no binding system. Two electrons or two positrons repel each other. It seems, therefore, that the dynamics of charge are similar in nature to the dynamics of color. Electrons are free particles; quarks were free only for a short time during the creation period. Also, three quarks can form a color singlet (a proton, for instance) but three electrons form a slate of charge -3, which obviously is not a singlet. The color property of quarks is often called "color quantum number" as a notation for the three color indices. Experiments have verified the existence of the color quantum number and, as we will see later, the color property is the source, together with gluons, of the strong interaction between quarks expressed in the QCD theory of formation of hadron particles such as protons, neutrons, and many others, all baryons or mesons (Figure 6.5).

Quarks and gluons were compressed in the fireball to enormous densities, so close that the strong nuclear force was too weak to attract their attention. As soon as the plasma expanded sufficiently, densities dropped considerably and the quarks came within a distance of 10^{-13} cm of each other, at which time the carriers of the strong nuclear force reacted, resulting in the permanent confinement of three quarks and many gluons in a predetermined, orderly way in the protons and neutrons (Figure 6.6).

As explained earlier, quarks appear in nature in three colors or three different charges. The u (up) quark has three colors identified for convenience as red, blue, and green and the same holds for the d (down) quarks—the only flavors or types of quarks that form permanent or stable particles, i.e., the proton and neutron. All

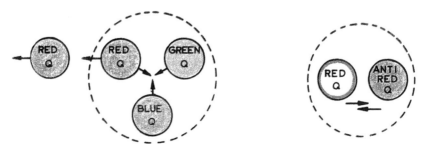

Figure 6.5. For baryons, three different quark colors attract, like colors repel. For mesons, color and opposite color attract (left, baryon; right, meson).

Quantum Chromodynamics

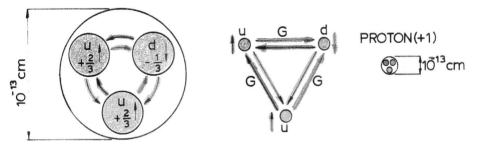

Figure 6.6. Schematic diagram of quarks in a proton and the strong nuclear force holding the three quarks confined, acting through gluons. (The sketch on the far right shows approximately a 4×10^{13} enlargement; the proton is actually 40 trillion times smaller than shown.)

other types or flavors of quarks, namely the c (charm), s (strange), b (bottom), and t (top) quarks, combine into hadrons in a similar way, but the hadrons are short-lived and decay into stable particles. We will discuss this later.

The Pauli exclusion principle forbids two identical spin $\frac{1}{2}$ particles (in the case of a proton the two u quarks) from occupying the same state and spin. Consequently, only the following combinations of u quarks can occur in protons: red–green, red–blue, green–blue.

In the case of a red–green combination, the d quark is blue. It is green if the u combination is red–blue. The effect of these color combinations is that protons, as all other hadrons, are color singlets. The spin $\frac{1}{2}$ of the proton relates to another law and to how nature controls the confinement of quarks in a hadron. If two quarks have the same flavor (u, d), they must spin in parallel. The third quark, if different, can spin parallel or antiparallel as is the case in the proton. As a consequence, the proton's spin is $\frac{1}{2}$ ($+\frac{1}{2} + \frac{1}{2} - \frac{1}{2} = +\frac{1}{2}$).

6.2. WHAT BINDS THE QUARKS TOGETHER?

1. The three different color quarks in the proton attract each other by means of the strong nuclear force.

2. The forces are coupled by gluons. Gluons are the particles that hold the universe together. Gluons are bosons, like photons, have no electrical charge, are massless, have spin 1, carry 50% of the momentum of a proton in flight, and have energy.

Gluons exert a color-dependent force that acts with different strengths on each of the three quarks' colors. A gluon, contrary to a photon, can change the color of a quark. For example, a gluon can transform a red quark into a blue quark by emitting a red–blue gluon (Figure 6.7).

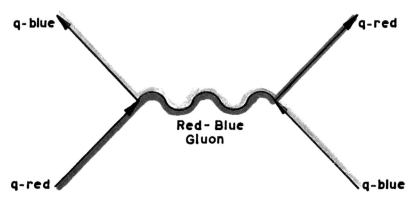

Figure 6.7. Transformation of a red (charge) quark to a blue (charge) quark by emitting a red–blue gluon.

Every gluon has a color and an anticolor (red, antired). The gluons are identified by their color-carrying properties. Three different colors allow nine different ways of coupling gluons to quarks—nine independent combinations of color and anticolor:

>red–green
>red–blue
>green–red
>green–blue
>blue–red
>blue–green
>red–red
>green–green
>blue–blue

The superposition or combination RR + BB + GG is completely symmetric and cannot change the color of any quark. This leaves eight different color couplings and eight different gluons. Gluons can be considered as lines carrying color indices and, contrary to photons, they not only interact with quarks but also between themselves. A "red–blue" (RB) gluon converts into a "red–green" (RG) gluon by emitting a "green–minus blue" gluon (Figure 6.8). A red quark turns into a blue quark by emitting a "red–minus blue" (R–B) gluon.

6.3. THE STRENGTH OF THE STRONG NUCLEAR FORCE

The strong nuclear force, nature's strongest force, represents an enormous concentration of force. It is comparable to the energy (work) required to lift 1 ton

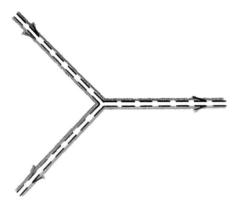

Figure 6.8. Transformation of a red–blue gluon into a red–green gluon, emitting a green–minus blue gluon.

of matter by 1 m, but it is acting on an extremely small area (10^{-13} cm) and small particles of matter such as quarks or protons.

The force weakens at very small distances and gives quarks a so-called asymptotic freedom at distances smaller than 10^{-13} cm and freedom at energy levels of 10^{15} GeV or larger. On the other hand, the force rises to its maximum at distances of 10^{-13} cm and permanently confines the quarks and gluons into hadrons. This phenomenon is called infrared slavery. The relative coupling strength of the strong force constant is

$$\alpha_{QCD} = 0.2\text{--}1$$

compared to

$$\alpha_{QED} = 1/137 = 0.0073$$

for the strength of electrodynamic interactions. The weak force is approximately 10^{-24} and the gravitational force 10^{-40} smaller.

6.4. THE NEUTRON

During the period of formation of protons after the big bang and considerable expansion and cooling of the universe, electrically neutral, free neutrons with a mass of 0.9396 GeV, slightly heavier than protons, were also formed. One u quark with charge $+\frac{2}{3}$ and two d quarks with charges of $-\frac{1}{3}$ combined, forming neutrons with spin $\frac{1}{2}$ (Figure 6.9). The two d quarks spin in parallel, the u quark antiparallel. Again, three different color charges combined into a color singlet.

Free neutrons are not stable and decay into protons, electrons, and anti-

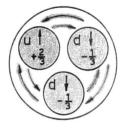

Figure 6.9. Schematic diagram of a neutron (spin $\frac{1}{2}$) consisting of two d quarks and one u quark. Charge: $\frac{2}{3} + (-\frac{1}{3}) + (-\frac{1}{3}) = 0$. (This sketch shows approximately a 4×10^{14} enlargement.)

neutrinos (n $\rightarrow$ p + e$^-$ + $\bar{\nu}_e$). The half-life of neutrons is approximately 15 minutes. The final ratio between neutrons and protons was 1:10 at the end of the quark–gluon nucleon production period during the expansion of the primordial fireball, which will be described in detail later.

6.5. EXPERIMENTAL PROOF OF THE QUARK–GLUON STRUCTURE OF PROTONS AND NEUTRONS

The structure of the proton and neutron was discovered in several experiments conducted first at the Stanford Nuclear Center (SLAC) in California. When high-energy electrons penetrated into protons, they either passed through or scattered off the quarks. The interaction of electrons with the quarks is determined by the electrical charges of the quarks. The angle of the electron scattering determined the charge of the quark. Charges of $+\frac{2}{3}$ and $-\frac{1}{3}$ have been confirmed for u and d quarks, respectively, in these and many subsequent experiments, as shown in Figure 6.10.

Another experiment, conducted with protons moving at close to the speed

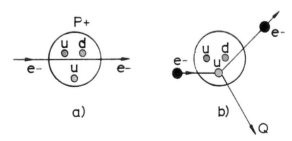

Figure 6.10. Experimental proof for quark–gluon structure of protons. Panel b demonstrates the scattering of an electron by a u quark.

Quantum Chromodynamics

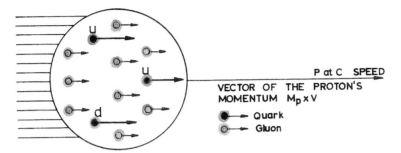

Figure 6.11. Schema of the momentum of a proton moving at close to the speed of light. More than 50% of the momentum is carried by the gluons.

of light, confirmed that the three quarks carry equal amounts of momentum as the proton itself, in compliance with the conservation law of physics stating that the total momentum of a system must be equal to the momentum of all its components (Figure 6.11). The experiments demonstrated that the three quarks carried less than 50% of the proton's momentum and that the remaining part was contributed by gluons, carriers of the strong nuclear force. Gluons have color and energy.

6.6. THE REACTION TIME OF PARTICLE INTERACTION

The strong force and its carrier, the gluons, are responsible for confining quarks into baryons and mesons. According to my theory of creation, quarks were born free during the creation process from pure electromagnetic energy. The strong nuclear force interacts not only with enormous force but with amazing speed. The quarks, after birth, traveled at velocities close to the speed of light. Since the size of the particles is less than 10^{-13} cm, the reaction time is less than 10^{-23} sec. Other interactions are much slower.

Electromagnetic reactions at 10^{-21} sec take 100 times longer, and decay processes of short-lived baryons and mesons sometimes take up to 10^{-9} sec, although certain decay processes are as fast as 10^{-23} sec. The time for birth, transition, and decay of the unstable particles may be as fast as 10^{-18} sec. Although the reasons are unknown, nature eliminates "unwanted" particles of matter as quickly as they are created and leaves only the few types of stable particles such as electrons, protons, neutrons, and neutrinos; neutrons only if they combine with protons into cores of atoms within 10–15 minutes after birth. Otherwise, neutrons as free particles decay into protons, electrons, and electron-antineutrinos. Nature's police force causing most of the decay is the weak nuclear force. Decays, however, are also caused by the electromagnetic force, and at very high energy levels by a substantial weakening of the strong force.

7

Conservation Laws

Earlier we briefly mentioned some of the important conservation laws prevailing in the cosmos. For clarity and to establish a coordinated view, we will review them in relation to the strong nuclear force interactions of quarks and gluons as they apply to the formation of particles.

Most conservation laws such as energy and momentum have general applications in the universe, while others govern exclusively the behavior and interaction of elementary particles. Every conservation law is connected with some symmetry in the laws of nature. For example, a particle moving freely through space conserves its momentum (mass × velocity) as long as space is homogeneous. If an obstacle is put in its path and space becomes inhomogeneous, the momentum of the particle will change and will no longer be conserved. The conservation rules enable the prediction of which particle transformations and decays can take place.

The conservation laws can be subdivided into three groups based on their physical origin. The first group is connected with the geometry of four-dimensional space-time and covers:

- The conservation of energy E, which is the result of the homogeneity of time.
- The conservation of momentum p, which is the result of the homogeneity of space. The three-dimensional space of the universe has been found to be homogeneous and isotropic or identical in all directions.
- The conservation of angular momentum M, which is the result of the isotropic property of space.
- The center-of-mass conservation law, which is the result of the equivalence of all inertial reference frames in four-dimensional space.

The second group combined all of the charge conservation laws. Five charges are additive characteristics of particles and all are conserved. The quantum nature or the origin of the charges is unknown. The five charges are:

- Q The electrical charge
- B The baryonic charge for all baryons
- L The leptonic charge for the electron and electron-neutrino
- L' The second leptonic charge for the muon and muon-neutrino
- L'' The third leptonic charge for the tau and tau-neutrino

The third group includes conservation laws that are valid only for certain particle interactions. They are:

- S Strangeness, for particles containing strange quarks (s). Particles that do not contain s quarks have $S = 0$.
- C Charm, for particles containing charm quarks (c). Particles that do not contain c quarks have $C = 0$.
- J Isotopic spin of particles.
- J_z Isotopic spin projection to the z axis.
- P The parity conservation law.

A discussion of the various conservation laws in each group follows.

7.1. GROUP 1

7.1.1. The Energy (E) Conservation Law

This law states that the total energy in all physical processes such as particle annihilation or creation of new particles is conserved. In other words, the sum of the energies of the final particles must equal the energies of the initial particles.

We take, as an example, a free neutron n, which decays within 15 minutes into a proton, electron, and electron-antineutrino. This decay caused by the weak interaction is called radioactive decay:

$$n \rightarrow p + e^- + \bar{\nu}_e$$
$$E_n = E_p + E_e + E_{\bar{\nu}_e}$$

The energy sum of the mass-energy of the proton, electron, and electron-antineutrino must equal the original energy of the free decaying neutron.

The relativistic equation for energy is

$$E_{rel} = \frac{Mc^2}{[1 - (v^2/c^2)]^{1/2}} \tag{7.1}$$

where v is the particle's velocity, M is the mass of the particle, and c is the speed of light. For $v = 0$ or for a particle at rest, $E_0 = Mc^2$. This is called the energy of the

Conservation Laws

rest mass or the energy equivalent to the mass of a given particle. The total energy of a particle in motion is the sum of the rest mass E_0 and the kinetic energy of its motion E_{kin}:

$$E_{rel\ total} = E_0 + E_{kin}$$
$$E_{kin} = E_{rel} - E_0 = E_{rel} - Mv^2$$

The kinetic energy vanishes when the particle is at rest or $v = 0$. As the speed approaches the speed of light, the mass of a particle increases considerably as shown in Figure 4.1.

7.1.2. The Momentum p Conservation Law

$$p = \frac{Mv}{[1 - (v^2/c^2)]^{1/2}} \tag{7.2}$$

The law of conservation of momentum ($M \times v$ = mass $\times$ velocity) states that the total momentum after collision of two particles such as protons or electrons will be the same as before the collision. If we take as an example the collision of two particles of equal mass M and equal speed v, traveling in opposite directions, their total momentum is 0 (Figure 7.1). After collision $[(Mv) + (-Mv) = 0]$, the total momentum of the two particles will also be 0. Both may fly after the collision in different directions and new particles may be created of total mass M_1 and M_2 of different velocities, as shown in Figure 7.2. However, their momenta (M_1v_1, M_2v_2) must be equal. All interactions of particles comply with the combined laws of energy and momentum.

7.1.3. The Angular Momentum Conservation Law or Conservation Law of Spin J

Most particles in the universe have angular momentum. They spin around their own axis of rotation, which is perpendicular to the particles' momentum. The spin of a free particle does not change in free motion and can be an integer 0, 1, 2, ... or a half integer $\frac{1}{2}, \frac{3}{2}, \frac{5}{2}$ ... $\times \hbar$ (Planck's constant = 1.054×10^{27} g/sec = 2.612×10^{-66} cm^2).

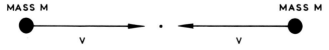

Figure 7.1. Total momentum of two particles of equal mass M and equal speed v, traveling in opposite directions, is 0.

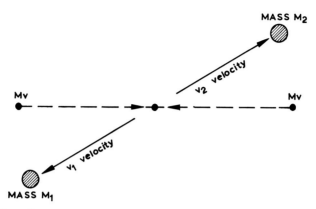

Figure 7.2. After collision of particles $(Mv) + (-Mv) = 0$, new particles may be created but $M_1v_1 = M_2v_2$.

This applies as well to quarks inside of baryons and mesons. The quarks occasionally orbit around each other in particles at excited, high-energy levels.

Massless particles such as photons (spin 1) can spin only parallel or antiparallel (left or right) to the direction of motion. Particles with mass can also spin perpendicularly to the direction of their motion.

7.1.4. J_z Spin Projection to the z Axis

7.1.5. The Center-of-Mass Conservation Law

The last two conservation laws are specific cases of major conservation laws described in detail above.

7.2. GROUP 2

7.2.1. The Law of Conservation of Electrical Charge Q

This law states that the total charge at the beginning of an interaction equals the total charge at the end. An electron cannot for instance be transformed into a neutrino or gamma ray (γ). These particles have no electrical charge and the law of conservation of charge for the electron (-1) would be violated. The electron would have to decay into a charged particle with smaller rest mass. The negative charge (-1) of an electron is equal in magnitude to the charge of a positron ($+1$) or proton ($+1$), but of opposite sign.

7.2.2. The Conservation Law of Baryon Charge B

By definition, the baryon number B for a proton or neutron is 1, as for other short-lived baryons. Mesons and all leptons have $B = 0$ and all quarks have $B = \frac{1}{3}$ (Three quarks = $3 \times \frac{1}{3}$ = baryon number 1.)

The total baryon number before and after interaction must be conserved. This law guarantees the stability of the proton. The conservation laws of energy, momentum, and charge would not prevent the decay of a proton into a positron and γ rays:

$$p \rightarrow e^+ + \gamma$$
$$Q = \phantom{p \rightarrow{}} +1 \phantom{{}+{}} +1$$

However, the baryon number would be violated:

$$p \rightarrow e^+ + \gamma$$
$$B = \phantom{p \rightarrow{}} 1 \phantom{{}+{}} 0 + 0$$

Also, the fact that the proton is the lightest particle carrying baryon charge 1 guarantees it against decay.

An antiproton has a negative baryon number, -1. In a proton–antiproton annihilation experiment (Figure 7.3), we start with a system for which $B = 0$. After the annihilation, the γ rays conserve the baryon number $B = 0$ and the particles born from the energy transfer to mass must still conserve the baryon number and maintain $B = 0$.

7.2.3. The Conservation Law of Lepton Numbers L, L', L"

Electrons, muons, taus, neutrinos, and their antiparticles are governed by conservation laws prohibiting certain interactions that other conservation laws would allow. The lepton number L for electrons is (1) and for positrons (-1), the muon number L' for muons is (1) and for antimuons (-1), and the tau number L'' for taus is (1) and for antitaus (-1). Before and after interaction, the numbers L, L', L'' must be maintained.

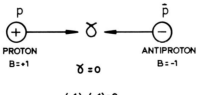

Figure 7.3. The baryon number is maintained after collision and annihilation of a proton (+) and antiproton (−). The γ electromagnetic energy has $B = 0$.

7.3. GROUP 3

7.3.1. Strangeness (S) Conservation Law

Baryons called hyperons (containing u, d, and s quarks) or mesons called kaons (which contain one u, d, or s quark and an antiquark) have strangeness $S = 1$. All baryons that do not contain the strange (s) quark (e.g., the proton and neutron) have zero strangeness, $S = 0$. The conservation of strangeness is maintained in strong and electromagnetic interactions. It is not always conserved in weak interactions. Strange particles are produced from adequate energy, after annihilation of stable, colliding particles, and always in pairs.

For example, a collision of two protons

$$p + p \rightarrow p + \Lambda^0 + K^+$$

produces a hyperon or lambda particle Λ^0 containing u, d, and s quarks and a meson K^+ containing a u quark and $\bar{s}$ antiquark (Figure 7.4). The Λ^0 baryon has electrical charge $Q = 0$, baryon number $B = +1$, strangeness $S = -1$, and a rest mass of 1116 MeV. The K^+ meson called kaon contains a u quark and $\bar{s}$ antiquark. K^+ has $Q = +1$ resulting from the $+\frac{2}{3}$ charge of the u quark and $+\frac{1}{3}$ charge of the $\bar{s}$ antiquark, $S = 1$, $B = 0$, and a rest mass of 498 MeV.

All charges are conserved in this process:

$$p + p \rightarrow p + \Lambda^0 + K^+$$

$B = 2$	$1 + 1$	$1 + 1 + 0$	$B = 2$
$S = 0$	$0 + 0$	$0 + -1 + 1$	$S = 0$
$Q = 2$	$1 + 1$	$1 + 0 + 1$	$Q = 2$

If only the Λ^0 particle were produced, the strangeness of the final system would be -1 instead of 0.

Strangeness is not conserved, however, in the decay of particles containing s

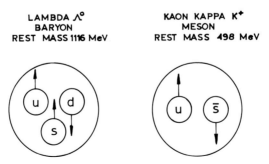

Figure 7.4. The collision of two high-energy protons in an accelerator can produce a baryon Λ^0 and a meson kaon K^+.

Conservation Laws

quarks. If it were, the Λ^0 particle could not decay and it would be stable like a proton. However, the Λ^0 particle decays in 2.5×10^{-10} sec into

$$\Lambda^0 \to p + \pi^-$$

or

$$\Lambda^0 \to n + \pi^0$$
$$S = \quad 1 \quad\quad 0 + 0$$

Strangeness in this weak-force decay is violated. The strangeness disappeared and the final particles do not contain the s quark, which transformed during the decay into a d quark and, occasionally, also into a c quark. It is also not conserved in particles containing, for instance, c and b quarks, which decay into s quarks.

7.3.2. Hypercharge Y

Strangeness is often combined with the baryon number and defined as hypercharge Y:

$$Y = S + B$$

The rest energy or mass, the hypercharge, and the spin describe the geometric pattern of a particle.

7.3.3 Charm (C) Conservation Law

Just like strangeness, charm is an additive, integral quantity and applies only to particles containing the charm quark c. It is conserved only in strong and electromagnetic interactions.

There are many particles, all nonstable, containing c quarks. One baryon, named lambda Λ_c^+, contains one u, one d, and one c quark. It has charm $C = 1$ and an electrical charge

$$Q = 1 \quad (u = \tfrac{2}{3} + d = -\tfrac{1}{3} + c = +\tfrac{2}{3} = +1)$$

There are several mesons containing c quarks and c antiquarks. All charm particles are born from energy in pairs in order to conserve the charm similar to particles containing quarks.

For example, in the collision and annihilation of an electron and positron, two D^0 mesons are born. A D^0 meson contains a c quark and a ū antiquark (ūc). Both the charm and electrical charge are conserved:

$$e^+ + e^- \to D^0 + D^0$$
$$\begin{array}{llll} C = 0 & 0 + 0 & 0\ (+1) + (-1) & C = 0 \\ Q = 0 & +1 + -1 & 0 \quad 0 & Q = 0 \end{array}$$

The charm conservation law is violated in weak interactions. A charmed meson D^0, for instance, decays into a kaon K^+ and a pion π^-, both mesons:

$$D^0 \rightarrow K^+ + \pi^-$$
$$C = 1 \quad\quad 1 \quad\quad 0 + 0 \quad\quad C = 0$$

The charm of all charmed particles disappears during their decay and the c quark is transformed into a u or d quark.

7.3.4. Spin *J*

Most elementary and nuclear particles spin around their own axes. This includes electrons, protons, neutrons, muons, taus, neutrinos, photons, quarks, and most of quark combination particles such as baryons and mesons. Rotation or spinning is a generic characteristic in the universe: all stars, planets, and galaxies rotate and many orbit as well.

The rotation feature extends from the macroscopic to the microscopic world of elementary particles of matter. The spin or angular momentum, which is a quantum vector quantity, has direction and magnitude and is a fundamental property of elementary particles describing their rotation.

Particles spin either in opposite directions (antiparallel) or in the same direction (parallel) and can also orbit around each other (Figure 7.5). Baryons consisting of the same three quarks can have different rest masses (weights) depending on the direction of the spin of the individual quarks, spinning parallel or antiparallel. A good example is two baryons containing the same u, d, and s quarks:

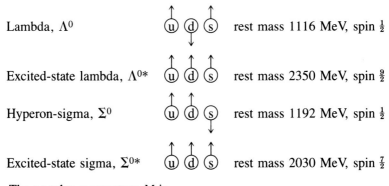

The angular momentum *M* is

$$M = \hbar J \tag{7.3}$$

where $\hbar$ is Planck's constant and *J* is either an integral or half-integral number $(0, \frac{1}{2}, 1, \frac{3}{2}, 2, \ldots)$. In the case of electrons, positrons, neutrons, and other composite baryons, the value of *M* is

Conservation Laws

Figure 7.5. Particles spin antiparallel or parallel and can also orbit around each other.

$$M = +\hbar \times \tfrac{1}{2}$$

or

$$M = -\hbar \times \tfrac{1}{2}$$

The minus sign describes the opposite spin direction. Planck's constant is usually set at 1 and all angular momenta are therefore multiples of $\hbar$.

The $+\tfrac{1}{2}\hbar$ and $-\tfrac{1}{2}\hbar$ spins become spins of $+\tfrac{1}{2}$ and $-\tfrac{1}{2}$ at the ground or lowest energy state. The same particle, however, can have, at higher energy levels, higher spins of $\tfrac{3}{2}, \tfrac{5}{2}, \tfrac{7}{2}$, etc. Sigma in the excited state (Σ^{0*}) has a spin of $\tfrac{3}{2}$ instead of $\tfrac{1}{2}$ at the ground state. Many mesons such as π^+, π^-, π^0 have 0 spin or no angular momentum, and photons have a spin 1. For mesons, however, the spin is always integral (0, 1, 2, . . .). The direction of the spin of quarks has a profound influence on the rest mass of a particle, as explained in the above examples with baryons Λ^0 and Σ^0, containing u, d, and s quarks.

Another good example is the (rho) meson ρ^+ consisting of ($\rho^+ = \bar{d}u$), a u quark and a $\bar{d}$ antiquark. It is approximately 610 MeV heavier than the meson having the same quark combination $\pi^+ = \bar{d}u$ but with parallel spin (Figure 7.6). It takes more energy to align the spins in the same direction and hence the larger rest mass or weight of those particles.

A delta Δ^+ baryon consisting of the same uud quark combination as a proton, but with all quarks spinning in parallel, weighs 30% more (Figure 7.7).

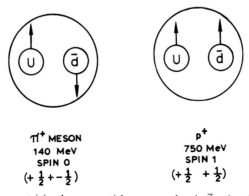

Figure 7.6. Two mesons containing the same particles—a u quark and a $\bar{d}$ antiquark—have substantially different weights due to the different spin of the particles.

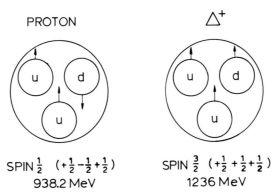

Figure 7.7. Although the Δ⁺ particle and proton have the same quarks, the former weighs 30% more because its quarks spin in the same direction.

7.3.5 Parity P

Another conserved quantity is P or intrinsic parity. Parity is conserved when nature does not distinguish between left and right. In strong interactions, such symmetry is observed and parity $P = +1$ or $P = -1$ can be assigned to all strongly interacting particles. In weak interactions, parity is violated; nature does not distinguish between left and right. For a proton $P = +1$; for a π meson, $P = -1$.

7.3.6. Isotopic Spin I_s

This is another conserved quantity in strong interactions. It has nothing to do with spin or angular momentum of the particle. What happens is that particles are grouped in multiples that have the same spin and baryon number but different charges. One example is the nucleon, which exists in two charge states: the proton, positively charged, and the neutron, neutral. These two states behave similarly. They are related by the isotopic spin. In mesons there is a typical charge multiple called the pion triplet: π^+, π^-, π^0.

The number for the multiplet M is related to the quantum number 1 by the equation

$$M = 2I_s + 1$$

For a nucleon, $I_s = \tfrac{1}{2}$; thus, $M_n = 2 \times \tfrac{1}{2} + 1 = 2$. For a pion, $I_s = 1$; thus, $M_p = 2 \times 1 + 1 = 3$.

8

The Particle Zoo

The known elementary and nuclear particles that can be stable are: quarks, protons, electrons, photons antiprotons, positrons, electron-neutrino, muon-neutrino, and tau-neutrino. Nevertheless, there are many quark–gluon composite particles in the form of mesons and baryons born from concentrated electromagnetic energy or γ photons, created in high-energy collisions of protons and electrons with their counterparts—antiprotons and positrons—in particle accelerators or in the upper atmosphere of the earth from collisions of particles with cosmic radiation from the sun, stars, and galaxies.

The birth of particles from pure energy is a relativistic effect of the well-known Einstein mass–energy equivalence formula, $E = Mc^2$. Various combinations between three quarks from the six known types, namely, u, d, s, c, b, and t, form short-lived baryons, all heavier than the proton. One of any of the six types of quarks together with an antiquark combine to form short-lived particles known as mesons. The subsequent fast decay of the created particles is caused in most cases by the weak nuclear force. The short-lived particles decay directly or indirectly into stable particles and/or in some cases back into electromagnetic energy or γ photons. The shortest possible lifetime of the particles, or the transit time, is 10^{-23} sec. It is the time required for light to cross the infinitesimally small space of a particle. Many particles have lifetimes considerably longer than the transit time. The charged meson called pion π^+, for instance, has a lifetime of 10^{-6} sec and other mesons called hyperons and kaons may live up to 10^{-10} sec.

Hundreds of particle variations have been discovered, since 1950, in cosmic radiation and accelerator laboratories. All are grouped into two categories—mesons and baryons—and are created from energy in proton–proton, proton–electron, proton–antiproton, and electron–positron collisions at high energies and velocities close to the speed of light. They all decay shortly after creation into protons, electrons, and neutrinos.

Mesons consist of one quark and one antiquark and decay very shortly after creation into electrons, positrons, neutrinos, and photons.

8.1. BARYONS

Baryons consist of three quarks held together by the strong nuclear force, are heavier than the proton, and have noninteger spin in units of $\hbar$, $\frac{1}{2}$, $\frac{3}{2}$, etc. All baryons except the proton decay; what remains are neutrinos, electrons, protons, and photons. Once the existence and characteristics of the six types (flavors) of quarks (u, d, s, c, b, and t) have been established, one could predict the various composite systems or quark configurations of baryons. No particles containing the t quark have so far been discovered. Due to its large rest mass, energy levels higher than those presently available will have to be achieved in particle accelerators.

The u and d quark combinations in protons and neutrons are the building blocks of the universe together with electrons, neutrinos, and photons. Other combinations of u and d quarks have been predicted and discovered, all guided by the rules of the various conservation laws and all short-lived. Spin direction of the quarks has a profound influence on the weight of the particles. Some of the major quark combinations are shown below (for complete classifications, see Tables 8.1 and 8.2).

$$\text{Spin} \rightarrow \begin{array}{ccc} u & u & u \\ \uparrow & \uparrow & \uparrow \end{array} = \Delta^{++}$$

$$\begin{array}{ccc} u & u & d \\ \uparrow & \uparrow & \uparrow \end{array} = \Delta^{+}$$

$$\begin{array}{ccc} u & d & d \\ \downarrow & \downarrow & \downarrow \end{array} = \Delta^{0} \qquad \text{Delta particles}$$

$$\begin{array}{ccc} d & d & d \\ \downarrow & \downarrow & \downarrow \end{array} = \Delta^{-}$$

Next, combinations of the u and d quarks with the strange (s) quark are called strange baryons or hyperons:

$$\text{Spin} \rightarrow \begin{array}{ccc} u & d & s \\ \uparrow & \downarrow & \uparrow \end{array} = \Lambda^{0} \quad \text{(lambda)}$$

$$\begin{array}{ccc} u & d & s \\ \uparrow & \uparrow & \downarrow \end{array} = \Sigma^{0} \quad \text{(sigma}^{0})$$

$$\begin{array}{ccc} u & u & s \\ \uparrow & \uparrow & \downarrow \end{array} = \Sigma^{+} \quad \text{(sigma}^{+}) \qquad \text{Hyperons}$$

$$\begin{array}{ccc} d & d & s \\ \downarrow & \downarrow & \uparrow \end{array} = \Sigma^{-} \quad \text{(sigma}^{-})$$

The Particle Zoo

Another system is called cascade hyperons:

$$\text{Spin} \to \begin{array}{ccc} u & s & s \\ \uparrow & \downarrow & \uparrow \end{array} \quad \Xi^0 \quad (xi^0)$$

$$\begin{array}{ccc} d & s & s \\ \downarrow & \uparrow & \uparrow \end{array} \quad \Xi^- \quad (xi^-) \qquad \text{Cascade hyperons}$$

$$\begin{array}{ccc} s & s & s \\ \uparrow & \uparrow & \uparrow \end{array} \quad \Omega^- \quad (\text{omega})$$

Very few baryons containing the charm (c) quark or charmed baryon have been discovered. The (lambda c) Λ^+_c baryon containing the u, d, and c quarks

$$\begin{array}{ccc} u & d & c \\ \uparrow & \downarrow & \uparrow \end{array}$$

is one of the heaviest baryons known, with a rest mass of 2273 MeV.

To date, no baryons containing the t quark have been discovered. However, a variety of mesons with the c quark and mesons containing the b quark have been found.

8.2. MESONS

Mesons of practically all predictable combinations of u, d, s, and c quarks have been recorded in particle collisions, with some containing the b quark. Mesons appear in groups. As with all particles, the spin direction of quarks and antiquarks has a substantial influence on the weight or rest mass of mesons.

The combinations of u and d quarks and their antiquarks occur in groups of three. The pions have an antiparallel quark–antiquark spin:

$$\begin{array}{cc} \bar{d} & u \\ \uparrow & \downarrow \end{array} = \pi^+ \text{ (pion}^+\text{)}$$

$$\begin{array}{cc} \bar{u} & d \\ \uparrow & \downarrow \end{array} = \pi^- \text{ (pion}^-\text{)}$$

$$\begin{array}{cccc} \bar{u} & u & / & \bar{d} & d \\ \downarrow & \uparrow & & \downarrow & \uparrow \end{array} = \pi^0 \text{ (pion}^0\text{)} \ (50\% \text{ of time } \bar{u}u; \ 50\% \ \bar{d}d \text{ system})$$

The rho mesons ρ^+, ρ^-, ρ^0 have parallel quark–antiquark spins of the same quarks and antiquarks and have substantially larger rest masses:

$$\begin{array}{cc} \bar{d} & u \\ \uparrow & \uparrow \end{array} = \rho^+$$

$$\begin{array}{cc} \bar{u} & d \\ \uparrow & \uparrow \end{array} = \rho^-$$

$$\begin{array}{cccc} \bar{u} & u & / & \bar{d} & d \\ \uparrow & \uparrow & & \uparrow & \uparrow \end{array} = \rho^0 \ (50\% \text{ of time } \bar{u}u; \ 50\% \ \bar{d}d \text{ system})$$

Table 8.1. Classification of Baryons

Quark mix	Category	Particle name	Notation Particle	Notation Anti-particle	Quark structure and spin	Mass (MeV)	Electrical charge Q	Baryon number B	Strangeness S	Charm C	Spin J	Half-lifetime (sec)	Decay particles
Quarks u + d	Nucleon	Proton	p	$\bar{p}$	u u d →←→	938.2	±1	±1	0	0	±½	Stable	—
		Neutron	n	$\bar{n}$	u d d →←→	939.6	0	±1	0	0	±½	0.93×10^3	$p + e + \bar{\nu}_e$
	Delta particles	Δ^{++}	Δ^{++}	$\bar{\Delta}^{++}$	u u u →→→	1232	2	±1	0	0	±3/2		$p + \pi^+$
		Δ^+	Δ^+	$\bar{\Delta}^+$	u u d →→←	1236	+1	±1	0	0	±3/2		$p + \pi^{++}$
		Δ^0	Δ^0	$\bar{\Delta}^0$	u d d →→←	1236	0	±1	0	0	±3/2		$p + \pi^0$
		Δ^-	Δ^-	$\bar{\Delta}^-$	d d d →→→	1241	−1	±1	0	0	±3/2		$N + \pi^-$
		Δ^{+**}	Excited state		u u d →→←	2420	1	±1	0	0	+1½	10^{-23}	$n + \pi$
	Lambda	Λ^0	Λ^0	$\bar{\Lambda}^0$	u d s →←←	1116	0	±1	±1	0	±½	2.5×10^{-10}	$p + \pi^-$ $p + \pi^0$
		Λ^{0**}	Excited state		u d s →←←	2350	0	±1	±1	0	±3/2	0.8×10^{-12}	$p + \pi^-$ $p + \pi^0$

			Quarks u d s	Mass						Mean life	Decay
Hyperons–sigma	Σ⁺	$\overline{\Sigma}^+$	u u s ↑→←	1189	±1	±1	±1	0	±½	0.8×10^{-10}	$p + \pi^0$
										10^{-14}	$n + \pi^+$
	Σ⁰	$\overline{\Sigma}^0$	u d s ←→←	1192	0	±1	±1	0	±½		$\Lambda + \gamma$
	Σ⁻	$\overline{\Sigma}^-$	d d s ←→←	1197	∓1	±1	±1	0	±½	1.5×10^{-10}	$n + \pi^-$
	Σ⁺*	Excited state	u u s ←→←	2030	+1	+1	+1	0	+7/2	10^{-23}	$\Lambda + \pi$
											$\Lambda(1520) + \pi$
Cascade hyperons	Ξ⁰	$\overline{\Xi}^0$	u s s ←→←	1315	0	±1	±½	0	±½	3×10^{-10}	$\Lambda + \pi^0$
	Ξ⁻	$\overline{\Xi}^-$	d s s ←→←	1321	∓1	±1	∓2	0	±½	1.7×10^{-10}	$\Lambda + \pi^-$
	Ξ⁻*	Excited state	d s s ←→←	1820	∓1	∓1	∓2	0	±3/2	10^{-23}	$\Lambda + \overline{K}$
	Ξ⁰*	Excited state	u s s ←→←	1530	0	∓1	−2	0	±3/2	10^{-23}	$\Lambda + \pi^0$
Omega	Ω⁻	$\overline{\Omega}^-$	s s s ←→←	1672	∓1	∓1	∓3	0	±3/2	1.3×10^{-10}	$\Xi + \pi$
											$\Lambda + K^-$
Lambda	Λ_c^+	$\overline{\Lambda}_c^+$	u d c ←→←	2273	±1	±1	0	1	±½		$\Lambda \pi^+ \pi^+ \pi^-$ or $pK^- \pi^+ u$

Table 8.2. Classification of Mesons

Particle name	Notation Particle	Notation Antiparticle	Quark structure and spin	Mass (MeV)	Electrical charge Q	Baryon number B	Strangeness S	Charm C	Spin J	Half-lifetime (sec)	Decay particles
Pion	π^+	π^-	$\bar{d}\,u$	140	± 1	0	0	0	0	2.6×10^{-8}	$\mu^+ + \nu_\mu$
Pion	π^-	π^+	$\bar{u}\,d$	140	∓ 1	0	0	0	0	2.6×10^{-8}	$\mu^- + \nu_\mu$
Pion	π^0	π^0	$\bar{u}\,u / \bar{d}\,d$	135	0	0	0	0	0	0.76×10^{-16}	$\gamma + \gamma$
Rho	ρ^+	ρ^-	$\bar{d}\,u$	750	∓ 1	0	0	0	1	10^{-23}	
Rho	ρ^-	ρ^+	$\bar{u}\,d$	750	∓ 1	0	0	0	1	10^{-23}	
Rho	ρ^0	ρ^0	$\bar{u}\,u / \bar{d}\,d$	750	0	0	0	0	1	10^{-23}	
Kaon kappa	K^+	K^-	$\bar{s}\,u$	498	± 1	0	± 1	0	1	1.2×10^{-8}	$\mu^+ + \nu_\mu$ $\pi^+ + \pi^0$
Kaon kappa	K^-	K^-	$\bar{u}\,s$	494	∓ 1	0	∓ 1	0	1	1.2×10^{-8}	$\mu^- + \nu_\mu$ $\pi^- + \pi^0$
Kaon kappa	K^0	$\bar{K}^0$	$\bar{d}\,s$	498	0	0	± 1	0	1	0.86×10^{-10}	$\pi^+ + \pi^-$

The Particle Zoo

					Mass (MeV)						Lifetime (s)	Decay modes
Phi	ϕ^0	ϕ^0	s	$\bar{s}$	1020	0	0	0	0	1	10^{-22}	$K^- + K^+$
	η'	η'	s	$\bar{s}$	958	0	0	0	0	0	10^{-22}	π^+, π^-, π^0; 2γ
	D^+	D^+	$\bar{d}$	c	1868	±1	0	±1	0	0		$\pi^+\pi^- + \pi^0$; $K^-\pi^+\pi^+$
	D^0	D^0	$\bar{u}$	c	1863	0	0	±1	0	0		$K^0 + \pi^+\pi^+$; $K^-+\pi^+\pi^+$
	F^+	F^+	$\bar{s}$	c	2040	±1	±1	0	0	0	10^{-23}	$K^- + \pi^+ + \pi^0$
(J-psi)	J/ψ	J/ψ	c	$\bar{c}$	3097	0	0	0	±2	1	10^{-20}	$\gamma + \gamma$
	B^-	B^+	c	$\bar{c}$	548	0	0	0	±2	1	2.4×10^{-19}	$\pi^+\pi^- \pi^0$
Excited state	J/ψ'	$J/\bar{\psi}'$	c	$\bar{c}$	3684	0	0	0	2	1	10^{-23}	$e^+ + e^-$
	B^-	B^+	b	$\bar{u}$	5260							
	B^0	$B^=$	b	$\bar{d}$	5260							
Upsilon	γ	$\bar{\gamma}$	b	$\bar{b}$	9460	0	0			1		$\mu + \bar{\mu}$
Excited	γ'	$\bar{\gamma}'$	b	$\bar{b}$	10000							$e^- + e^+$
Excited	γ''	$\bar{\gamma}''$	b	$\bar{b}$	10400							

Mesons containing combinations of u or d and s quarks and antiquarks have also been discovered:

$$\bar{d}\uparrow \; s\downarrow = K^0 \text{ (kaon}^0)$$

$$u\uparrow \; \bar{s}\downarrow = K^+ \text{ (kaon}^+)$$

$$\bar{u}\downarrow \; s\uparrow = K^- \text{ (kaon}^-)$$

$$s\uparrow \; \bar{s}\uparrow = \phi^0 \text{ (phi)}$$

All predictable combinations of charm c quark with u, d, and s quarks have also been recorded:

$$c\downarrow \; \bar{u}\uparrow = D^0$$

$$c\downarrow \; \bar{d}\uparrow = D^+$$

$$\bar{c}\uparrow \; s\uparrow = F^+$$

$$\bar{c}\uparrow \; c\uparrow = J/\psi \text{ (angular momentum } J = 1)$$

$$\bar{c}\uparrow \; c\downarrow = \eta_c \text{ (angular momentum } J = 0)$$

The first discovered combinations of u, d, and b quarks are:

$$u\uparrow \; b\downarrow = B^-$$

$$d\uparrow \; b\downarrow = B^0$$

$$b\uparrow \; b\downarrow = \Upsilon \text{ (upsilon)}$$

The hunt is on for other possible variations of baryons and mesons, especially for those containing the heavy quarks b and t. As soon as more powerful accelerators now under construction become available, higher energy levels will be achieved and heavy particle combinations with b and t quarks will be discovered. The basic baryon and meson particles and their characteristics are given in Tables 8.1 and 8.2.

8.3. EXCITED STATES OF PARTICLES

In many ways, baryons and mesons are similar to atoms. Atoms are composite structures with a positively charged nucleon core of protons and neutrons held by the strong force and orbiting electrons with negative charge, and held together with the core by the electromagnetic force.

The spinning rates of the nucleons and electrons and the rate of orbiting electrons create various excited forms of atoms. Similarly, the total spin of a nuclear particle is the result of the spin of individual quarks. If there is orbital motion, then the total spin is larger and the same applies to the mass of the particles. To excite a quark, however, requires much more energy than in the case of an electron orbiting in an atom, as the strong force between the quarks provides resistance to excitation.

Baryons consisting of three $\frac{1}{2}$-spin quarks have, in the ground state, a spin $\frac{1}{2}$ when one quark rotates antiparallel ↑ ↑ ↓ ($\frac{1}{2} + \frac{1}{2} - \frac{1}{2} = \frac{1}{2}$). If all quarks rotate in parallel ↑ ↑ ↑ ($\frac{1}{2} + \frac{1}{2} + \frac{1}{2} = \frac{3}{2}$), the baryon has spin $\frac{3}{2}$. In excited states and with orbiting quarks, the spin of baryons can reach a high level of 15/2.

8.4. BARYON ISOSPIN MULTIPLETS

Between the many basic baryons, there are six groups that have the same spin and baryon number and differ only in their electrical charge. The groups are called isospin multiplets:

- Omega Ω
- Cascade hyperons Ξ
- Sigma Σ
- Lambda Λ
- Nucleons (protons and neutrons)
- Delta Δ

Most of the particles shown in Tables 8.1 and 8.2 are those with the lowest energy level or ground state and rest mass. In other words, the constituent quarks have the lowest spin and no orbital movement. The quarks do not orbit around each other. A proton, for instance, in a ground state energy level has a spin of $\frac{1}{2}$. In an excited state where additional energy is applied, the proton may have a much larger spin of $\frac{3}{2}$, $\frac{5}{2}$, $\frac{7}{2}$, etc., and a larger mass. The higher energy particles are called recurrences of low-lying states. The proton's rest mass of 938.2 MeV with a corresponding spin of $\frac{1}{2}$ may rise to 1512 MeV with spin $\frac{3}{2}$ or even to 1688 MeV with spin $\frac{5}{2}$.

Mesons, in their ground or lowest energy level, have spin 0 or 1. Nonparallel rotation of the quarks ($\frac{1}{2} - \frac{1}{2} = 0$) gives the particles a spin 0.

Parallel rotating quarks ($\frac{1}{2} + \frac{1}{2} = 1$) give the particles a spin 1. In excited states with orbiting quarks, the spins of mesons go up to 4.

8.5. MESON ISOSPIN MULTIPLETS

In a similar manner as for baryons, there are meson isospin multiplets: kappa, pions, and rho.

A few examples of excited state particle occurrences are shown in Figure 8.1. In the ground-state D_0 and $D^\pm$ mesons, the c quarks and their "partner" antiquarks $\bar{d}$ in D_0 and $\bar{u}$ in D^+ spin in opposite directions, resulting in particles having spin $J = 0$ and a mass of 1863 and 1866 MeV, respectively. In the excited form, the quarks and antiquarks spin in the same direction, resulting in a much higher rest mass (weight) of around 2000 MeV and spin $J = 1$. The excited-state particles return rapidly (in approximately 10^{-23} sec) to the ground state, emitting pions π or γ rays.

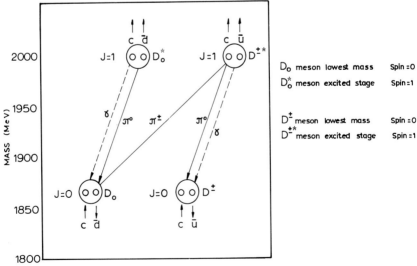

Figure 8.1. The excited states of D mesons.

The Particle Zoo

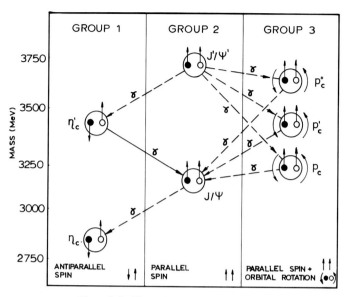

Figure 8.2. The excited states of J/ψ mesons.

8.6. THE EXCITED STATES OF THE J/ψ MESONS

The particle called the J/ψ meson and its excited states (Figure 8.2) are possibly the best example of the influence of spin level and direction, as well as orbital movement on the rest mass. All of the seven mesons, consisting of a charmed quark c and its antiquark $\bar{c}$, have different combinations of spin energy and orbital angular momentum. Let us analyze the diagram and its three groups.

8.6.1 Group 1

When the quarks' spin is antiparallel, two η mesons are formed—η_c and η'_c. This state is called *paracharmonium*, similar to the electron e^- and positron e^+ pair called *positronium* in electromagnetism. These particles are the result of decay from two excited states of J/ψ. As η_c has a spin $J = 0$ (↑ ↓), it cannot be created directly from pure energy in the form of γ photons with spin 1. This would violate the conservation law of spin, which states that the final system must have the same total spin value—1.

8.6.2. Group 2

When the quarks' spin is parallel, the $c\bar{c}$ quark combination forms the well-known J/ψ particle and its higher-energy-level "brother" J'/ψ' particle. The spin of both J/ψ and J'/ψ' is $J = 1$. The particle can therefore be created through an electron–positron annihilation into pure photon energy. This system of J/ψ and J'/ψ' is called *orthocharmonium*.

8.6.3. Group 3

When the quarks rotate in parallel and enter into orbital rotation with an angular momentum of 1, a new meson particle known as p_c is created in three states of energy or rest mass (weight) ranging between the weights (energy) of J/ψ and J'/ψ'. This state in physics is called a p-wave charmonium, with quarks spinning in the same direction and orbiting around each other as a bi-star system.

Similarly to excited atoms, when excited electrons in higher orbits drop to normal orbits and photons are emitted, particles at higher energy levels transform into lower energy systems by emitting in this case high-energy photons.

There are many other similar particles comprising the same type of quarks, existing for a short period of time in different levels of energy excitation and then decaying and emitting photons in the form of high-frequency γ rays to lower levels. They all finally decay to stable leptons and protons.

A good example is the Λ^{0*} baryon (uds) recurrence of mass 1815 MeV and spin $\frac{5}{2}$ instead of the ground-state particle of mass 1116 MeV and spin $\frac{1}{2}$, or Λ^{0**} of mass 2350 MeV and spin $\frac{9}{2}$. Also, the Δ^{++} baryon (uuu) (ground state mass 1236 MeV and spin $\frac{3}{2}$), when excited to a higher spin (11/2), reaches a mass of 2420 MeV.

9

The Weak Nuclear Force

As we have seen, the strong nuclear force is responsible for the formation of quark-containing nuclear particles of matter such as baryons and mesons. However, all of the particles we have analyzed—created in reactions due to cosmic radiation in the upper earth's atmosphere, but mainly in high-energy particle accelerators—decay shortly after creation, except for the proton (which is stable) and the neutron only as part of the nucleus of atoms. Free neutrons decay in approximately 15 minutes into a proton, electron, and electron-antineutrino.

The weak nuclear force, which does not seem to have a creative mission in the universe, is responsible mainly for the decay of unstable particles. It initiates and polices the rapid and orderly decay of unwanted particles of matter and reestablishes quickly and efficiently nature's preference for simplicity and uniformity. At the end of the decay process of unstable particles, all that remains are the stable particles of the universe: protons, electrons, neutrinos, and photons.

The first manifestation of the weak force was the 1896 discovery by Becquerel of the radioactive β-decay of uranium nuclei. Neutrons in the nuclear core of uranium atoms radiate so-called β-particles (high-energy electrons). This is a natural phenomenon of radioactive decay caused by the weak force interaction.

In 1933 when Fermi developed the first theory of the weak force and Pauli predicted the neutrino ν and its counterparticle antineutrino $\bar{\nu}$ to explain the imbalance in energy and momentum in transmutation of neutrons into protons, the protons and neutrons at that time were considered to be elementary particles. The internal quark structure theory of the nucleons was developed only 30 years later, independently by Murray Gell-Mann and George Zweig.

The present explanation for the decay of neutrons and other hadrons is the interaction of the weak force mediated by particles called bosons ($W^{\pm}$ and Z^0).

They fall into a similar category as gluons mediating the strong interactions and photons mediating the electromagnetic interactions.

What seemed at the time of Fermi to be a simple decay process of β-radiation of the neutron decay resulting in:

$$\underset{\text{(neutron)}}{n} \rightarrow \underset{\text{(proton)}}{p} + \underset{\text{(electron)}}{e^-} + \underset{\text{(electron-antineutrino)}}{\bar{\nu}_e}$$

is explained today as the decay or transmutation of the d quark in the neutron via the weak interaction into a u quark, electron, and antineutrino (Figure 9.1). Similar decays take place with charmed particles of baryons and mesons (Figure 9.2). The charm (c) quark in both reactions is transmuted into a strange (s) quark by weak reactions mediated by the boson W⁻ creating at the same time an antimuon $\bar{\mu}^+$ and muon-neutrino ν_μ.

Similarly, these interactions altering the electrical charge (n = 0 charge, p = +1) fall into one of two types of the weak force interaction, in this case the *charged-current interaction*. There is another process that does not change the electrical charge as is the case in proton–neutrino scattering:

$$\nu_e + p \rightarrow \nu_e + p$$

This weak force interaction is called *neutral-current interaction*. The incoming electron-neutrino emits a virtual Z⁰ boson, which reacts with the proton (Figure 9.3). In all of these processes, the four particles with spin ½ are called fermions.

How is this explained? First, let us determine *the strength of the weak force*. Fermi established the strength of the weak force or G_F based on the mass of the proton, which is $m_p \cong 1$ GeV. The Fermi constant is a parameter with the dimension of (energy)⁻²:

$$G_F \simeq \frac{1.16 \times 10^{-5}}{m_p^2} = 1.16 \times 10^{-5} \text{ GeV}^{-2}$$

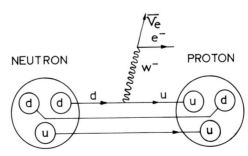

Figure 9.1. An incoming neutron emits a virtual boson W⁻ and turns into a proton. The W⁻ boson disintegrates into an electron and electron-antineutrino. A d quark decays into a u quark.

The Weak Nuclear Force

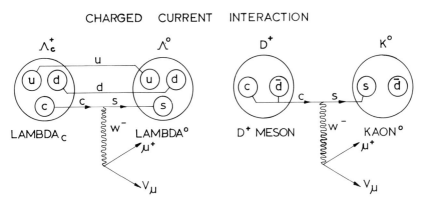

Figure 9.2. The charm (c) quark in both reactions in the baryon Λ_c and meson D^+ is transmuted into a strange ρ quark via the weak force. The boson W^- decays into a muon and muon-neutrino.

In comparison, the electromagnetic force is a dimensionless quantity:

$$\alpha = 1/137 \simeq 10^{-2}$$

For neutron β-decay at 1 GeV energy-mass, the relation is

$$\frac{G_F}{\alpha} \simeq \frac{1.16 \times 10^{-5}}{10^{-2}} = 1.16 \times 10^{-3} = \frac{1.16}{1000}$$

The weak force is approximately 1000 times weaker than the electromagnetic force at low energy.

As mentioned, the charged weak-force processes such as β-decay are mediated by positive and negative bosons W^+ and W^-. Z^0 bosons mediate the neutral weak-process scattering of a neutrino and proton, for example. W and Z

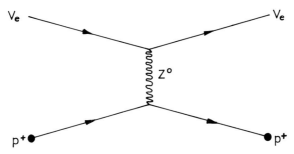

Figure 9.3. The incoming electron-neutrino emits a neutral boson of the weak interaction Z^0. Neutral-current interaction does not change the charge of the particles.

Table 9.1. Carriers of the Weak Force

Bosons	Spin	Mass (GeV)	Electrical charge
W+	1	8.3	+1
W−	1	8.3	−1
Z⁰	1	9.0	0

bosons were discovered in 1983 in actual tests at CERN and their rest mass came close to predicted value (Table 9.1).

It is really not meaningful to compare the electromagnetic constant α with the weak-force G_F Fermi constant, which is a number with the dimension of GeV^{-2}. What is relevant is to compare α with the strength of W and Z particles:

$$G_F = \frac{\pi}{\sqrt{2}} \frac{\alpha}{M_w^2}$$

where M_w is the mass of the W boson = 8.3 GeV.

The strength of the weak and electromagnetic forces are comparable. It appears that weak interaction and electromagnetism are different manifestations of the same force, which is called the electroweak process.

Interactions of W⁻ bosons with quarks and leptons are shown in Figure 9.4. In a similar way, the W⁺ boson mediates the transmutation of a neutrino into an electron and a u quark into a d quark (Figure 9.5).

The heavier leptons muon (μ) and tau (τ) can also be turned into neutrinos ν_μ and ν_τ as well as a c quark into an s quark—all mediated by W⁻ and W⁰ bosons.

Leptons and quarks seem to act as equivalent groups:

$$\begin{pmatrix} e^- \\ \nu_e \end{pmatrix} \quad \begin{pmatrix} u \\ d \end{pmatrix}$$
$$\begin{pmatrix} \mu^- \\ \nu_\mu \end{pmatrix} \quad \begin{pmatrix} c \\ s \end{pmatrix}$$
$$\begin{pmatrix} \tau^- \\ \nu_\tau \end{pmatrix} \quad \begin{pmatrix} t \\ b \end{pmatrix}$$

The Interaction between Particles

An analysis has been presented of the main characteristics, if not of all then certainly of the majority, of the elementary and nuclear particles. Let us review the different classes of particles to gain a better understanding of their interdependence with the four forces of nature—the strong nuclear force, the electromag-

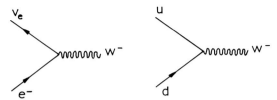

Figure 9.4. Interactions of W⁻ boson with electrons and quarks. The electron turns into a neutrino; the d quark turns into a u quark emitting a boson W⁻.

netic force, the weak nuclear force, and gravitation (see Tables 9.2 and 9.3). There are six distinct groups:

1. *Hadrons* participate in the strong nuclear interactions and have two subdivisions—baryons and mesons.
2. *Baryons* (e.g., proton, neutron) consist of three quarks, obey the Pauli exclusion principle, have nonintegral spin ($\frac{1}{2}$, $\frac{3}{2}$, . . .), and baryon number 1.
3. *Mesons* (e.g., π, K) consist of one quark and one antiquark, do not obey the Pauli principle, have integral spin (0, 1, 2), and baryon number 0.
4. *Leptons* are elementary particles. They are electrons, muons, taus, neutrinos, and their antiparticles. As far as we know, electrons and neutrinos are indivisible. They do not participate in the strong interactions, interact with the electromagnetic and weak nuclear force, and have spin $\frac{1}{2}$.
5. *Fermions* is the generic term for all particles with spin $\frac{1}{2}$, such as baryons and leptons.
6. *Quarks* are the basic building blocks of compounded particles of matter such as baryons and mesons. They existed as free particles only during the creation period of the universe at very high levels of energy, 10^{15} GeV and above, and equivalent temperatures of 10^{20} K and above. In the fireball, they were squeezed very closely together and the forces between them at short distances were extremely small. This phenomenon is called

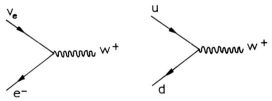

Figure 9.5. A W⁺ boson mediates the transmutation of a neutrino into an electron and a u quark into a d quark.

Table 9.2. The Four Forces of Nature

Name of interaction (force)	Main range (cm)	Relative intensity or strength	Strength parameter	Mediator Name	Mediator Mass	Mediator Baryon number	Mediator Spin	Particles participating in interaction
Strong nuclear	10^{-13}	1	QCD = 0.2–1	Gluon	0	0	1	All hadrons
Electromagnetic	∞	10^{-4}	$\frac{1}{137.036}$ (= 0.0073)	Photon	0	0	1	Leptons
Weak nuclear	10^{-15}	10^{-24}	1.16×10^{-8} GeV^{-2}	$W^{\pm}$ boson, Z^0 boson	8.3 GeV	0	1	Leptons, hadrons, decay
Gravitation	∞	10^{-40}	6.673×10^{-8} cm^3/g-sec^2	Graviton	9 GeV	0	2	All

The Weak Nuclear Force

Table 9.3. Elementary Particles and Forces of Interaction (Neglecting Gravity)

Particle name	Notation Particle	Notation Anti-particle	Stable	Not stable	Responds to the following interactions	Theory of interactions
Fermions (spin $\frac{1}{2}$)						
Leptons						
Electron (−) Positron (+)	e^-	e^+	√		Electromagnetic and weak	QED
Electron-neutrino	ν_e	$\bar{\nu}_e^-$	√		Only weak	
Muon	μ^-	μ^+		√	Electromagnetic Weak	QED
Muon-neutrino	ν_μ	$\bar{\nu}_\mu^-$	√		Only weak	
Tau	τ^-	τ^+		√	Electromagnetic Weak	QED
Tau-neutrino	ν_τ	$\bar{\nu}_\tau^-$	√		Only weak	
Quarks						
Up quark	u	$\bar{u}$	√	Can decay into d	Strong Electromagnetic Weak	QCD QED
Down quark	d	$\bar{d}$	√	Can decay into u	Strong Electromagnetic Weak	QCD QED
Strange quark	s	$\bar{s}$	√		Strong Electromagnetic Weak	QCD QED
Charm quark	c	$\bar{c}$	√		Strong Electromagnetic Weak	QCD QED
Bottom quark	b	$\bar{b}$	√	Can decay	Strong Electromagnetic Weak	QCD QED
Top quark	t	$\bar{t}$		Possibly decays	Strong Electromagnetic Weak	QCD QED
Photon	γ		√		Strong Electromagnetic Weak	QCD QED

The Three Types of Forces

Gauge symmetry	Force	Carriers–intermediate vector bosons
$SU^{(2)}$	Weak	W^+, W^-, Z^0 bosons
$U^{(1)}$	Electromagnetic	γ photon
$SU^{(3)}$	Strong	Eight gluons (g_1–g_8)

asymptotic freedom. Almost all particles are unstable in the free state except for the proton, electron, photon, antiproton, positron, three types of neutrino, and quarks (only at high temperature and energy levels at creation time).

A deep understanding of the laws of particle creation and interactions is essential to grasp the creation moment when, according to my theory of creation, primordial electromagnetic energy condensed into elementary particles such as quarks, electrons, and neutrinos as well as their antiparticles. Later, after the explosion of the primordial fireball, the high-temperature soup of the basic particles and electromagnetic energy in the form of γ photons, expanded and the individual forces of nature acted to combine the u and d quarks in protons and neutrons. When the temperature dropped sufficiently, the latter combined with electrons and created the basic elements of the universe—hydrogen and helium. All other elements were created later, in the cores of stars.

The above interpretation of particle creation of the universe will be described later in the Velan theory of creation.

We proceed in Chapter 10 with an analysis of the fourth force of nature—gravitation.

10

Gravitation

The theories of gravitation describe the force between two masses. Newton related the level of the gravitational force to the mass of two given objects and the distance between them. The gravitational force F_G between body M_1 and m_2 (Figure 10.1) at a distance d can be calculated from the Newtonian formula of gravitation:

$$F_G = \frac{GM_1 m_2}{d^2} \qquad (10.1)$$

where G is Newton's gravitational constant.

According to the Newtonian theory of gravitation, the gravitational force acts instantaneously and everything in the universe moves in a straight line, unless it is acted upon by some outside force. In accordance with the equation (10.1) of gravity, if the mass m_2 is electromagnetic energy in the form of photons (γ), which have zero mass, the gravitational force of mass M_1 and m_2 is

$$F_G = \frac{GM_1\, 0}{d^2} = 0$$

It is clear therefore that, in accordance with the Newtonian theory of gravitation, the force of gravity cannot influence light beams or any other form of electromagnetic energy. As we will see shortly, this has been shown by Einstein and subsequent experiments to be wrong. A light beam is bent when passing near a large body such as a star.

Einstein's special theory of relativity proved in the first instance that gravity does not act instantaneously over distance as shown in Figure 10.1. The classical field theory of Einstein proved beyond any doubt that nothing can move faster than the speed of light and, therefore, the gravitational field propagates force at finite speed as shown in Figure 10.2.

One of the main foundations of modern physics is Einstein's general theory

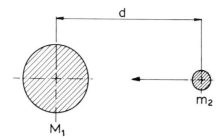

Figure 10.1. Newtonian gravity. Gravitational force F_G acts instantaneously over distance d.

of relativity, which relates the force of gravity to the structure of space-time. In this theory, space and time are unified in a four-dimensional continuum called space-time. Mass curves space-time and determines its curvature. Space-time and mass become two undivided companions in the universe. Mass shapes space-time and space-time controls the movement of mass through space-time. In the absence of mass or energy, space-time is flat and there is no gravity.

As shown schematically in Figure 10.3, space-time is curved near a massive object M_1 and the motion of particle m_2 is along a path that follows the path in curved space-time called a geodesic, not a straight "Newtonian" line. The particle itself exerts a reciprocal influence on space-time, causing gravitational waves that move at the speed of light and disturb the geodesic on which the particle moves.

10.1. EFFECTS OF GRAVITY ON ELECTROMAGNETIC ENERGY

As mass and energy are equivalent ($M = E/c^2$), gravity affects equally matter and energy in the form of electromagnetic radiation or even gravitational energy propagated by gravitational waves and gravitons. In other words, light or even gravitational energy is affected by the forces of gravity.

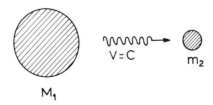

Figure 10.2. Classical Einstein theory of gravitation. The gravitational field propagates force with finite speed of maximum speed of light.

Gravitation

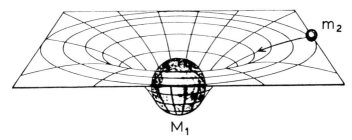

Figure 10.3. Einstein's general relativity concept of gravity. Massive object M_1 curves space-time. Mass m_2 moves on a geodesic—the shortest possible path.

Einstein, with his almost superhuman capability of discovering complicated physical laws of nature without the benefit of experimental proof at the time, predicted in his general theory of relativity both of the major effects of gravity on electromagnetic radiation such as light, namely:

1. The deflection of light passing near objects with large mass concentration curving nearby space-time (Figure 10.4).
2. Loss of starlight energy due to gravitation.

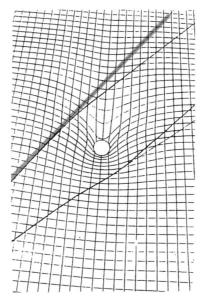

Figure 10.4. Light travels in a straight line (red line) in the absence of a gravitational field. Light is bent when passing through curved space.

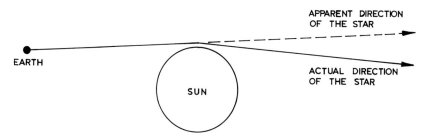

Figure 10.5. During the sun's total eclipse, starlight from a distant star has been deflected, moving the image of the star; the deflection is within 1% of the values predicted by Einstein's theory.

During the total eclipse of the sun on May 19, 1919, Einstein's gravitational theory, expressed in the general relativity theory, was proven for the first time (Figure 10.5). Many additional measurements taken regularly during eclipses, with more precise instrumentation, have given deflections within 1% of the values predicted by general relativity. Further dramatic proof for light bending is the so-called gravitational lens effect. A very large curvature of space-time induced by a massive galaxy can create two or more images of the same object, as shown schematically in Figure 10.6. In an actual space photograph (Figure 10.7), where a large galaxy acts as a gravitational lens, two quasar images appear.

The most accurate measurements of electromagnetic wave-bending by gravitation were achieved with radar signals sent to the Viking spacecraft located on Mars and bounced back to earth by the spacecraft transponders. Gravitation deflects and delays the electromagnetic signals. The closer the signals pass by the sun, the larger is the delay in the time of arrival on earth of the signal, compared to the straight path between earth and the Mars spacecraft that would exist in the absence of gravitation and curved space-time by the sun (Figure 10.8).

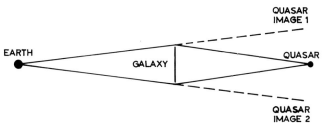

Figure 10.6. Schematic diagram of the gravitational lens effect by a large galaxy bending the quasar's light, giving two images.

Gravitation

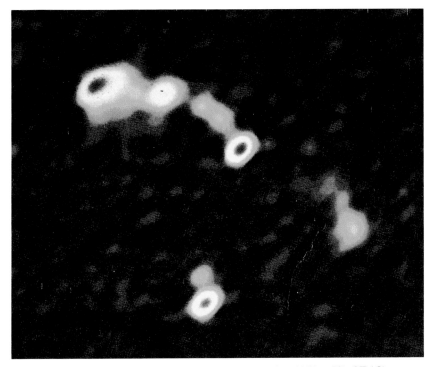

Figure 10.7. Actual NRAO image of the gravitational lens 0957 + 561. (NRAO)

10.2. THE GRAVITATIONAL FORCE

The Newtonian concept of the gravitational force acting on everything was replaced with the Einsteinian concept of curved space-time caused by the presence of mass. The classical Newtonian formulas of gravitation, however, still apply with good accuracy in areas of space-time with relatively small mass concentrations such as the earth or our solar system. The "force" of gravity F between masses M_1 and m_2 is $F_G = GM_1m_2/d^2$. The force felt by the larger object M_1 is the same as that felt by the smaller object m_2 (Figure 10.9). G is the gravitational constant = 6.673×10^{-8} cm^3/g-sec^2. The constant is very small—0.000000066. Two small objects of 1 g each separated by 1 cm attract each other with a force not more than 6.673×10^{-8} dyne or 6.673×10^{-11} g. For the hydrogen atom, the gravitational force between the proton and electron is given by the same formula, as shown in Figure 10.10.

The gravitational force is the weakest of the four forces of nature. The electromagnetic force between the proton and electron of a hydrogen atom is:

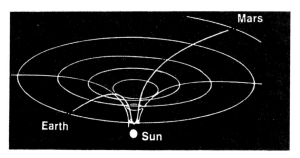

Figure 10.8. Path of radar bounced back to earth by the transponders on the Viking spacecraft on Mars. Signals are delayed due to curved space-time by the sun.

$$F_{\text{Epe}} = \frac{e^2}{r^2} \qquad (10.2)$$

where e is the electrical charge. The ratio between gravitational [Eq. (10.1)] electromagnetic forces [Eq. (10.2)] is

$$F_{\text{epe}}/F_{\text{Gpe}} \rightarrow G\frac{m_p m_e}{r^2} \bigg/ \frac{e^2}{r^2}$$

$$= G\frac{m_p m_e}{e^2} \simeq 10^{-40}$$

The ratio of these two forces is immense. Gravity is 10^{40} times weaker than the electromagnetic force and is therefore neglected in high-energy particle physics.

When mass becomes very large, however, there is a marked difference between the Newtonian theory and general relativity. As a rule, the difference becomes important when α is larger than a few hundredths of a percent:

$$\alpha = \frac{GM}{Rc^2} \qquad (10.3)$$

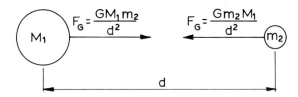

Figure 10.9. The gravitational force felt by both objects is the same.

Gravitation

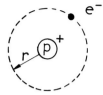

Figure 10.10. Gravitational force between the electron and proton is:

$$F_{Gpe} = \frac{Gm_p m_e}{r^2}$$

where G is the gravitational constant, M the mass, R the radius of the object, and c the speed of light.

The effects of general relativity in the solar system, for instance, are relatively small when compared to effects elsewhere in the universe. At the surface of the sun, for instance, α is only $2 \times 10^{-6}\%$, so the Newtonian theory can still be used well to calculate events on earth and to calculate orbits in the solarsystem. However, with astronomical objects such as galaxies, large stars, quasars, and other massive concentrations of matter where α is 0.01–2%, space is considerably curved and the Newtonian results must be corrected by the effects of general relativity.

10.3. EFFECT OF GRAVITY ON ENERGY

The second prediction made by Einstein of a gravitational redshift of light emitted in a strong gravitational star field is more difficult to measure. The difficulty is that the gravitational redshift effect is insignificant when compared to the Doppler shifts caused by the simultaneous receding velocity of a star, galaxy or the motion of gases in the same area. However, observations of gravitational redshift effects on light coming from white dwarfs—remnants of smaller stars with small diameters but large mass concentration—have determined unequivocally the slowdown and energy loss of light due to strong gravitation.

To recapitulate, general relativity or the theory of gravity teaches us that mass and energy curves space-time and shapes its geometry. In return, any movement of mass and energy changes the geometry of space-time. It is generally recognized that the created disturbance is propagated by gravitational waves, which travel at the speed of light and have a quantum energy in the form of gravitons with spin 2. So far, however, gravitational waves have not been detected on earth. The present instrumentation technology is not sensitive enough to detect the minute deformations the gravitational waves would cause in the detectors.

The intensity of gravitational waves or the amount of energy they carry is extremely small. For instance, the earth, orbiting around the sun, should emit gravitational waves with an energy of approximately 0.001 watt. This loss of energy would result in the earth "falling" toward the sun due to loss of gravitational energy—10^{-6} cm (1 millionth of a centimeter) in 1 billion years. The effect of gravitational waves during a supernova explosion, however, is 1 billion times larger.

The energy E_g of a single graviton or quantum of gravity can be expressed as

$$E_g = h\omega_g \tag{10.4}$$

It is the product of Planck's constant h and the frequency ω_g of gravitational waves.

10.4. GRAVITATIONAL EFFECTS EQUAL TO ACCELERATION

Another cornerstone of the general theory of relativity is the principle that gravitation is equivalent to acceleration. This is called the principle of equivalence. It implies that falling freely in a gravitational frame is equal to being deep in space, far from any masses and gravitation.

A good example of the equivalence principle is an astronaut in an orbiting spacecraft, exposed to near-zero gravity and experiencing weightlessness. During the descent to earth, when the spacecraft accelerates, he experiences heavy gravitation. To him it seems that the gravitational field has been suddenly switched on.

The concept of four-dimensional space-time and the principle of equivalence gave Einstein the basis for arriving at the simple theory of general relativity. The ten field equations that describe nature's most mysterious phenomenon— gravity—can be written in a combined form as:

$$G = 8\pi T \tag{10.5}$$

where G is the measure of the curvature of space-time and T is the measure of the stress-energy content of space-time, or the more familiar expression of mass and energy that produce curvature. The field equations describe the curvature created by the stress energy and how the latter induces curvature on space-time.

The equation governs the external space-time curvature, the generation of gravitational waves or ripples in the curvature of space by stress-energy in motion. It contains within the equation of motion—which is force = mass × acceleration—the matter whose stress-energy generates the curvature. In practical terms, the equation governs the motion of planets, the deflection of light by stars, the collapse of a star to form a black hole, the expansion of the entire universe and its eventual collapse.

11

Black Holes, Quasars

The ultimate triumph of the "weakest" force in the universe over matter and space-time is achieved in black holes. The theoretical and mathematical predictions of black holes in 1939 by Oppenheimer and Synder were finally indirectly proven 34 years later by observing a binary star system in Cygnus X-1, where a large companion star becomes a black hole and pulls considerable quantities of material from the companion star.

A black hole is the remnant of the total collapse of a massive star's core at the end of its life, when all thermonuclear reactions cease and the core of at least 2.5 solar masses can no longer resist its own gravitational forces.

In smaller cores up to 1.44 solar masses, the electrons, squeezed to a density of 1000 tons per cubic inch, become what is called degenerate gas and can resist the gravitational collapsing forces. The remnant of such a star becomes a white dwarf, 10,000 km in diameter.

In larger cores of up to 2.5 solar masses, electrons are squeezed into protons and the mass becomes a degenerate neutron gas squeezed to a density of 10 billion tons per cubic inch, which can resist further collapse, and the star's core becomes a neutron star with a diameter of approximately 20 km.

If the remaining core of a large star after a supernova explosion of the star's shell contains more than 2.5 solar masses, the degenerate neutron gas can no longer resist the increasing gravitational forces.

Present theories on black holes presented by leading cosmologists such as Stephen Hawking and Martin Rees explain that once the resistance of the degenerate neutron gas to gravity is overcome and matter is squeezed to a smaller and smaller volume until it is totally crushed to practically a single point called a singularity, the star's core becomes a black hole.

The singularity is for all practical purposes a mathematical point of 0-space with infinite pressure and density of the crushed matter and infinitely curved

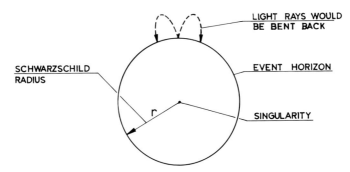

Figure 11.1. Schematic diagram of a black hole. The center point or singularity is a mathematical point with infinite density and infinite curvature of space. At the event horizon, gravity is so strong that even light cannot escape and is confined inside the horizon.

space-time. As to the contents of a singularity, there are several contradictory views. The theory of general relativity breaks down at this point and Einstein declared that his theory would not apply to these conditions of infinite levels of energy and temperature. All other laws of physics break down as well. We are describing here the generally accepted theory of black holes. My own theory is described in Chapter 24.

A black hole has an area around the center point or singularity called the event horizon (Figure 11.1). The event horizon's diameter depends on the total mass of the black hole. At the event horizon, the gravity is so strong that nothing can escape.

The gravity and curvature of space-time increases inside the event horizon until it becomes infinite at the singularity. The event horizon is a sphere and its diameter can be calculated from the Schwarzschild equation:

$$r = \frac{2GM}{c^2} \quad \text{cm} \tag{11.1}$$

where r is the radius (diameter = $2r$), G is the gravitational constant = 6.673×10^{-8} cm^3/g-sec^2, c is the speed of light = 2.997×10^{10} cm/sec, and M is the mass in grams of the collapsed object. For example, the event horizon of a 12-solar-mass black hole is 70 km in diameter. There is virtually nothing there except the victorious triumph of gravity having squeezed the mass out of existence, resulting in a highly warped region of space and time. All of the energy and mass of the black hole is tied up in the gravitational field. Black holes are invisible.

There are basically two major views as to the contents of a singularity:

1. There are those who claim that all matter and radiation have been squeezed into infinite density, creating infinite gravity and infinitely warped space-time around the black hole.
2. Others hold that a black hole is empty. All of the original matter of the star has been crushed out of existence and all that remains is infinitely warped space-time.

In both cases it is an incredible triumph of the weakest force in the universe—gravity—over matter, radiation, and even space-time.

The conservation law of energy can still be met if we consider that matter and radiating energy have been transformed into equal energy contained in the gravitational field and warped space-time, if this is possible.

There are frequent contradictions in the statements of the black hold theory specialists. One can read on one page in a specialized book on "Black Holes" that matter "has been completely crushed out of existence and all that exists is a region of infinitely warped space-time," all this in spite of the claims that the black hole maintains three characteristics—mass, spin, and charge. A few pages later, one can read that black holes contain matter, for without gravity that accompanies matter, a black hole could never form and that the mass of a black hole is an important characteristic.

11.1. SCHWARZSCHILD BLACK HOLE

This is the simplest type of a static black hole (Figure 11.2). It has no electrical charge or spin. It contains a point singularity in the center where all matter and radiation were crushed to infinite density. It has one static limit where particles cannot remain at rest, which is called the event horizon. Within the area of the event horizon, gravity is so strong that not even light can escape. Gravity

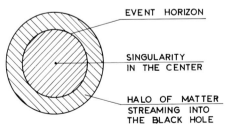

Figure 11.2. Schwarzschild black hole. It has no spin or electrical charge. It has a singularity in the center where matter is crushed to infinity and an event horizon.

and curvature of space increases toward the singularity where it is infinite. Time also stops inside the event horizon and for an observer just outside the event horizon stands at zero forever. It would also have a halo of matter streaming from companion stars into the black hole. As nearly all stars rotate, it is unlikely that this type of black hole exists in nature.

11.2. KERR ROTATING BLACK HOLE

Most massive cores of stars end up as Kerr black holes (Figure 11.3). They have a rapid spin but no electrical charge.

In the early stages of collapse, it has two event horizons and two ergospheres, which are elliptical areas around the horizons from which energy can be extracted. It has a ring-type singularity which lies in the horizontal line, perpendicular to the axis of rotation.

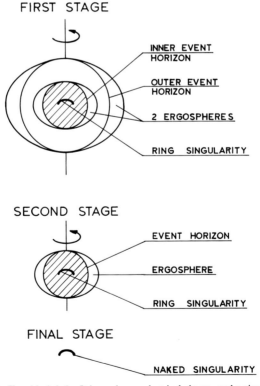

Figure 11.3. Kerr black hole. It has spin, no electrical charge, and a ring singularity.

As the radius of the collapsing mass reduces, the spin rapidly increases to maintain the conservation law of orbital momentum. As a result, the outer event horizon merges with the inner horizon. When the spin increases further, the inner horizon and ergosphere move toward the singularity and can finally vanish, leaving a naked singularity and infinite warped space. This type of black hole would be the expected variety for collapsed cores of massive stars.

11.3. REISSNER–NORDSTROM ELECTRICALLY CHARGED NONROTATING BLACK HOLE

This type has an electrical charge, no spin, and a pointlike singularity, just as the Schwarzschild type (Figure 11.4). It probably does not exist in nature as the electrical charge of particles would normally neutralize itself during the collapse.

11.4. KERR–NEWMAN ROTATING AND CHARGED BLACK HOLE

This type has both electrical charge and spin, ring-type singularity, and is similar in structure to the Kerr black hole (Figure 11.5). As the electrical charge of the collapsing mass would neutralize, such black holes probably do not exist in nature.

11.5. THE SEARCH FOR BLACK HOLES

It is obvious that black holes as such cannot be observed directly with telescopes, as nothing, not even light rays, can escape the overwhelming gravitational field and warped space around a black hole. However, indirectly, black holes can be discovered due to their immense gravitational field on a companion

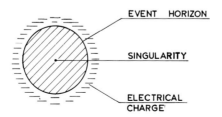

Figure 11.4. Reissner–Nordstrom black hole. It has charge, no spin, and a pointlike singularity.

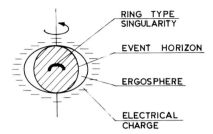

Figure 11.5. Kerr–Newman black hole. It has both charge and spin and a ring-type singularity.

star in a binary system or by causing the gravitational lens effect on passing light rays.

We have already seen the gravitational lens effect on the deflection of light coming from a quasar (Figure 10.6) as effected by a strong gravitational field of a large galaxy. A black hole, if located between the earth and a galaxy, can distort the view of the galaxy, as shown in Figure 11.6. We would see two distorted images.

Another way to indirectly detect black holes is to observe stellar wind from a companion binary star pouring matter into a disk around a black hole, which was originally the second binary star. This is shown in Figure 11.7. The candidate for the black hole in this case is the remnant of binary star Cygnus X-1, which drags matter from its companion star HDE 226868 with 20 solar masses.

The stellar matter from the visible giant star HDE 226868 pours into a huge disk around the black hole, pulled by the enormous gravity. The infalling gas is

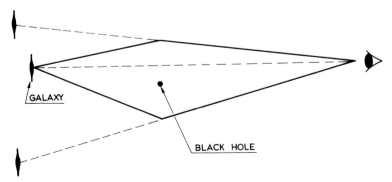

Figure 11.6. Black hole as gravitational lens. Light rays from a distant galaxy are deflected by a black hole's enormous gravity. We see two images of the galaxy.

Black Holes, Quasars 105

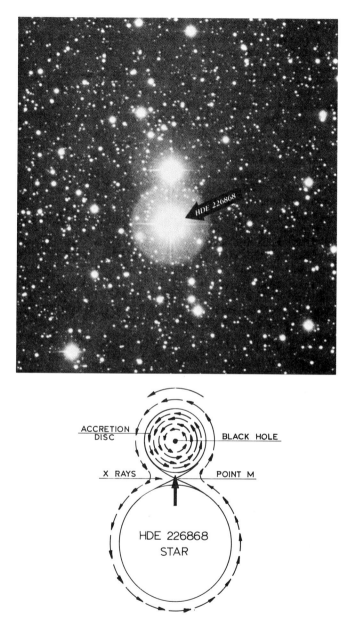

Figure 11.7. The Cygnus X-1 black hole. The photo shows the giant star HDE 226868, which is part of a binary system. The invisible companion became a black hole. The matter falling into the black hole is shown schematically in the diagram. (California Institute of Technology)

heated to highly elevated temperatures of millions of degrees Kelvin as it spirals into the black hole emitting X rays, and the accretion disk just above the black hole horizon becomes transparent, emitting also photons on the visual light wavelengths.

11.6. QUASARS—SUPERGALACTIC BLACK HOLES

Supermassive black holes have been discovered in the center of galaxies, some close to us such as the Exploding Galaxy NGC 5128 in Centaurus (Figure 11.8) 13 million light-years away from earth, the galaxy M82 (Figure 11.9) 10 million light-years away, or the Seyfert Galaxy M77 (Figure 11.10). Centers of these and perhaps most galaxies have powerful black holes—sources of enormous infrared, X-ray, radio-wave radiation—called quasars. But most powerful quasars were discovered at enormous distances of 13–15 billion light-years and speeding away at nearly $0.9c$ or 270,000 km/sec. Thus, when the universe was still young, approximately 3–4 billion years old, unusual events of cosmic

Figure 11.8. The Exploding Galaxy NGC 5128 in Centaurus. (Lick Observatory photograph)

Figure 11.9. The Peculiar Galaxy M82 (also called NGC 3034) in Ursa Major. (Lick Observatory photograph)

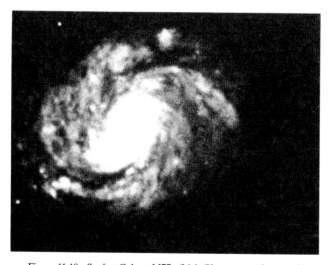

Figure 11.10. Seyfert Galaxy M77. (Lick Observatory photograph)

dimensions took place in the centers of some galaxies. Infalling gas of galactic magnitude as large as 5–8 billion solar masses collapsed in those galaxies in a runaway manner creating a supermassive black hole in the center and causing stars to crowd around the galaxy's center. Slow rotating stars have been swallowed by the black hole while fast rotating stars have survived.

If the universe is in a repetitive cycle, following the theory of rebounce, which I accept, there is an excellent explanation for the 15-billion-year-old quasars, provided you accept the Velan theory (presented later). Black holes, in the singularity model of creation, would be absorbed during the contraction cycle and become part of the singularity. In the Velan model, with a finite fireball, they will create a high local concentration of density and gravity and soon after the explosion of the fireball, the black holes will attract mass and radiation and early in the expansion cycle create quasars.

11.7. THE COSMIC DYNAMO PRINCIPLE OF A QUASAR

The rotation of the infalling accreting material confines it to the equatorial plane, perpendicular to the axis of rotation. In the same plane rotates a concentrated galactic magnetic field brought in by the massive infalling magnetized matter, generating a superpowerful electric field and creating a sort of cosmic dynamo. From the powerful radiation, electrons are created from virtual particles and shoot out from the galaxy at close to the speed of light. In addition, the infalling star-size masses are heated to hundreds of millions of degrees Kelvin when they spiral into the black hole emitting powerful gamma rays, together with radio waves, X rays, and light, resulting in phenomenal brightness. These quasars are not large, about 1 light-week, 200 billion km, or only 15 times larger than the diameter of the sun, but their luminosity or brightness is equal to 100–1000 galaxies of 100 billion stars each.

Some quasars speed away at 0.9 times the speed of light and though they are as far as 15 billion light-years away, they can be observed today with our largest telescopes, despite the fact that their activity culminated early in the evolution of the universe and subsided considerably at a later time.

As shown schematically in Figure 11.11, the gases of particles, highly magnetized, spiral toward the quasar creating enormously concentrated and rotating magnetic fields. As the concentrated magnetic field rotates down at enormous speed around the black hole, it creates a cosmic electric dynamo. It surrounds the black hole with a gigantic and powerful electric field and this energy is transformed into pairs of electrons and positrons, shooting out at nearly the speed of light along with radio waves, X rays, and infrared radiation.

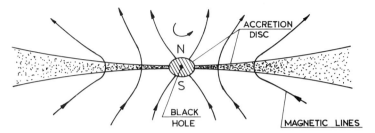

Figure 11.11. The cosmic dynamo principle of a quasar or supermassive black hole in the center of a galaxy.

11.8. THE SINGULARITY THEORY AS A PRIMORDIAL COSMIC BLACK HOLE

According to the standard or classical theory of cosmology, our universe originated from an explosion of a primordial singularity, seemingly containing all of the matter and radiating energy of the universe. If we were to consider such a singularity as a primordial cosmic black hole, we could calculate the Schwarzschild radius of the event horizon of such a singularity. This would be the space from which nothing, not even light, could escape. We arrive at startling results:

$$r_{ps} = \frac{2GM}{c^2}$$

Taking M as the critical mass of the universe ($= 5.68 \times 10^{56}$ g) and r_{ps} as the radius, we arrive at

$$r_{ps} = \frac{2 \times 6.685 \times 10^{-8} \times 5.68 \times 10^{56}}{9 \times 10^{20}} \quad \text{cm}$$

As 1 light-year $= 0.94605 \times 10^{18}$ cm,

$$r_{ps} = \frac{8.43 \times 10^{28}}{0.946 \times 10^{18}} = 8.92 \times 10^{10} \text{ light-years}$$

This result tells us that the event horizon of the primordial singularity before the big bang would extend to 90 billion light-years. The present radius is estimated to be 18 billion light-years and, at maximum expansion, approximately 35 billion light-years. So far, no one has explained why such a singularity, where even light cannot escape, would suddenly become unstable and explode—the big bang.

11.9. QUANTUM GRAVITY

Gravity, as we have seen, is completely different from the three other forces of nature. They all act in space-time, which serves as the background for the electromagnetic, weak, and strong nuclear force interactions. Gravity, on the other hand, interacts with space-time itself. According to Einstein's theory, a gravitational field is a curvature of space. While Einstein's theory of general relativity and its view of gravity has led to a well-defined and experimentally proven understanding of the universe and its evolution on the large macroscopic scale, it fails to explain events on the microscopic scale of 10^{-33} cm or even in the range of 10^{-17} cm (size of an electron) or 10^{-13} cm (size of a proton).

Correlating the two theories—quantum gravity and Einstein's theory of general relativity—is the problem, and presently there is no acceptable solution.

There is no acceptable theory of gravitation consistent with the parameters of quantum mechanics. In a simplified way we could probably unify the two theories by saying that particles of matter are pulled by virtual gravitons (quantum gravitation), similar to the encounter of two electrons, and the particles follow the curved trajectory of space-time predicted by Einstein's theory. One of the problems is that the quantum world is always moving, it is never still. The topography in the tiny Planck space-time cells (10^{-33} cm) fluctuates constantly, sometimes violently with total collapse and rebounce, virtual particles appear from the active vacuum of space influencing the three forces on the local scale. The curvature of space-time itself and its structure is subject to fluctuations.

Quantum gravity is still in many ways speculative and no satisfactory theory has been established, though many aspects of quantum mechanics have been proven experimentally in the presence of gravitational fields. The main problem with the singularity theory is that it is not renormalizable. For instance, it is not predictive because it requires an infinite number of experimentally determined constants.

In the quantum mechanics field theory, all four forces of nature interact with particles via bosons and so the "Newtonian gravitational force" is created between two particles by the exchange of massless particles with spin 2 called gravitons. So far gravitons have not been found experimentally.

A neutron, as a pointlike particle in classical physics, has mass and magnetic momentum. However, as discussed previously, it also has proven wavelike characteristics. A neutron wave acts in the same way as any other wave such as X rays or light. When two waves of equal amplitude meet, they interimpose. When they are exactly in phase, they interfere and the resulting amplitude is twice as great. When they are exactly out of phase, they interface destructively and cancel out.

The wavelength of neutrons at room temperature is 10^{-8} cm and the speed of propagation is 10^5 cm/sec, which is 10^{-5} times less than the speed of light. For comparison the wavelength of X rays, which are high-energy electromagnetic waves, is also typically 10^{-8} cm.

As mentioned earlier, the propagation of the neutron or any elementary particle waves follows the equation of Schrödinger:

$$\text{Wave length } \lambda = h/p$$

where h is Planck's constant and p is the momentum:

$$p = m_n \times v$$

where m_n is the mass of the neutron and v is its velocity of propagation. Consequently,

$$\lambda = \frac{h}{m_n \times v}$$

It has been proven experimentally in neutron interferometer experiments that the Schrödinger equation works well in the presence of gravity. And so here clearly we have the conflict between the two theories on the microscopic scale.

In quantum theory, all phase-dependent effects of gravitational fields depend on the mass of the particles through their wavelength (wavelength depends on the momentum, and momentum depends on the mass). In Einstein's theory of gravity, everything is based on geometry alone (curved space-time) in terms of position and momentum.

There are even greater discrepancies between the two theories when we analyze the gravitational effects in the tiny (10^{-33} cm) Planck space cells or superspace.

11.10. GRAVITY IN PLANCK SPACE (10^{-33} CM)

In Planck's theory of quantum mechanics, he established for the small space cells (10^{-33} cm), sometimes called superspace, their own quantities of length, time, and gravity, using three known constants: the gravitational constant G, the Planck constant $\hbar$, and the speed of light c. The Planck length L is:

$$L = \left(\frac{\hbar G}{c^3}\right)^{1/2} = 1.6 \times 10^{-33} \text{ cm} \qquad (11.2)$$

The Planck mass M is

$$M = \left(\frac{\hbar c}{G}\right)^{1/2} = 2.2 \times 10^{-5} \text{ g} \qquad (11.3)$$

This mass can be considered a Planck black hole. The Planck time T is

$$T = \left(\frac{\hbar G}{c^5}\right)^{1/2} = 5.3 \times 10^{-44} \text{ sec} \tag{11.4}$$

which is the time required for light to cross the Planck length of 1.6×10^{-33} cm. The Planck density is

$$\frac{M}{L^3} = \frac{c^5}{\hbar G^2} = 5.157 \times 10^{93} \text{ g/cm}^3 \tag{11.5}$$

As we will see later, this is the density of virtual (to be) particles in the vacuum of space, which become real particles in the presence of an external source of energy.

The gravitational energy created by the Planck mass of 2.2×10^{-5} g concentrated in a space with radius $L = 10^{-33}$ cm is enormous:

$$E_{\text{GRPL}} = \frac{G \times M_{\text{PL}}^2}{L} = \frac{6.673 \times 10^{-8} \times 4.84 \times 10^{-10}}{1.6 \times 10^{-33}} \tag{11.6}$$

Using 1 eV = 1.602×10^{-12} erg and 1 erg = 0.624×10^3 GeV,

$$E_{\text{GRPL}} = 20 \times 10^{15} \times 0.624 \times 10^3 = 12.48 \times 10^{18} \text{ GeV}$$

$$E_{\text{GRPL}} \cong 10^{19} \text{ GeV}$$

Thus, although the gravitational effect on small elementary particles is insignificant in the range of 10^{-13}–10^{-8} cm, it becomes enormous in superspace describing space cells in the range of 10^{-33} cm.

12

Unification of the Four Forces

Let us briefly review the properties of the four forces before we return to the unification attempts. The four forces of nature show a large variation in properties, their effective range and strength.

Range

The uncertainty principle of quantum mechanics determines the effective range or distance d for transmitting the forces by the equation:

$$d = \frac{\hbar}{mc} \quad \text{cm} \tag{12.1}$$

where $\hbar$ is Planck's constant, c is the speed of light, and m is the rest mass of the force-transmitting particle or boson.

Strong Force. The pion with a rest mass of $m_p \simeq 0.1$ GeV transmits the strong force between nucleons (protons + neutrons) in the nucleus of an atom; using Eq. (12.1), the effective distance d for the strong force is 10^{-12}–10^{-13} cm.

Weak force. The W (W^+, W^-), and Z^0 bosons transmit the weak force. They have a rest mass of $m \simeq 10^2$ GeV and therefore the effective range d of the weak force is 10^{-15} cm.

Electromagnetic Force. As the mass of the photon (γ) transmitting the electromagnetic force is zero, the effective range d is $\hbar/(0 \times c) = \infty$. The range for the electromagnetic force is infinite.

Gravitation. As gravitons transmitting gravitation have no mass, gravitation also has an infinite range.

Though the electromagnetic and gravitational forces have an infinite range, their influence declines as the square of the distance between the particles.

The strong force between quarks in hadrons or between nucleons in the nucleus of atoms is the most powerful. If we define the strong force interaction between two protons as 1, then the electromagnetic force is 10^{-2}, the weak force 10^{-15}, and the gravitational force 10^{-40}. For instance, if the electron of a hydrogen atom bound by electron attraction were to be bound by the gravitational force of the proton, the atom of hydrogen would be larger than the universe.

The interactions between particles caused by the various intermediary bosons can best be depicted by use of Feynman diagrams, shown in Figures 12.1–12.3.

When a neutron decays into a proton, an electron and an electron-antineutrino are formed:

$$n \to p + e^- + \bar{\nu}_e$$

What actually happens is shown in Figure 12.3. One d quark of the neutron (udd) decays into a u quark, creating a proton (uud). The virtual boson W^- emitted disintegrates at once into an electron and electron-antineutrino.

Examples of strong force interaction are shown in Figure 12.4 and gravitational interaction in Figure 12.5.

In spite of the diverse properties of the four forces, it is natural to search for the unification or common origin of all four forces.

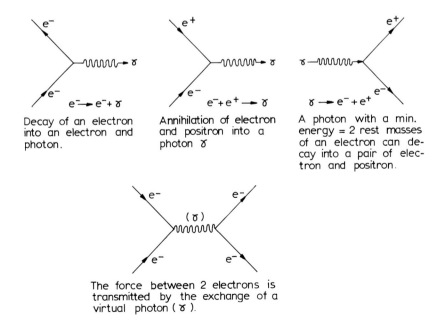

Figure 12.1. Electromagnetic interactions.

Unification of the Four Forces

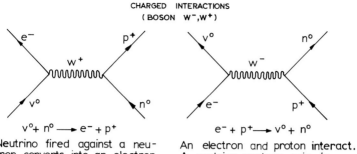

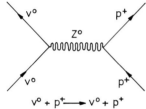

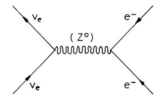

Figure 12.2. Weak force interactions.

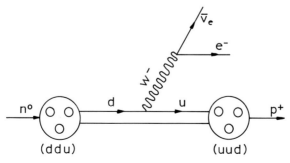

Figure 12.3. Decay of neutron. One d quark decays into a u quark and the emitted W⁻ bosons decay into an electron and electron-antineutrino.

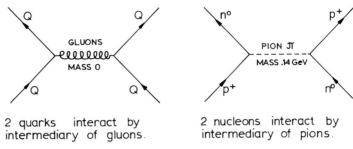

Figure 12.4. Strong force interactions.

The first success was achieved by James Clerk Maxwell when he unified electricity with magnetism into electromagnetism. Glashow, Weinberg, and Salam showed that electromagnetism and the weak force could be explained as two sections of the electroweak force.

The fundamental constant of quantum electrodynamics, also called the fine structure constant, is $\alpha = e^2/hc$, defined as the square of the charge of the electron, divided by the product of Planck's constant and the speed of light. It is equal to 1/137 ($= 0.0072992$) and it describes the strength of the electromagnetic force.

The strength of the weak interaction force is described by the Fermi constant, denoted G. Expressed in units of energy $G = 294$ GeV. Actually, $G/(hc)^3 = 1.166 \times 10^{-5}$ GeV^{-2}

$$(1/1.16 \times 10^{-5})^{1/2} = (86,200)^{1/2} = 294 \text{ GeV}$$

This indicates that the weak interactions are much weaker than the electromagnetic interactions. However, the figures indicate that at energies greater than 294 GeV, the standard picture of the weak interactions will not apply.

To see its full strength, we need to probe at small distances of 10^{-17} cm,

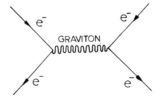

Figure 12.5. Gravitational interaction of matter and radiation through massless gravitons or quanta of gravitation.

Unification of the Four Forces

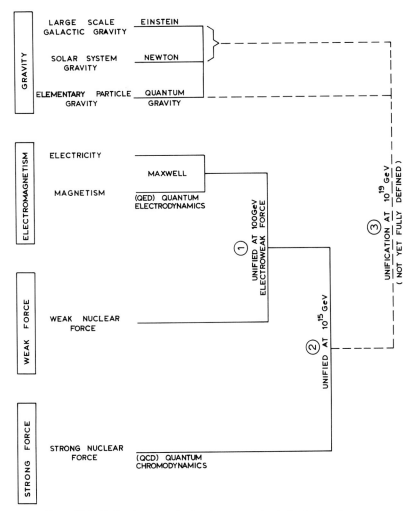

Figure 12.6. Unification attempts at various energy levels for the four forces.

which requires energies of over 100 GeV. At this level of energy, the weak force will be equal to the electromagnetic force and the two forces unite.

On the other hand, the strong force gets much weaker at high energy levels, to the point that at energies of 10^{15} GeV the strong force is so weak that quarks become free particles. Extrapolating the strength of the three forces, all three seem to be equal at energy levels of 10^{15} GeV.

The electromagnetic force would rise at this level of energy from 1/137 to approximately 1/42 and the strong force would fall to the same level of 1/42 from unity. Figure 12.6 and Table 12.1 outline the unification process.

In order to unify the three forces, leaving for the time being the gravitational force unrelated, we can group the particles, mainly the singlets electron and electron-neutrino and the color triplets quarks and antiquarks, in the following way:

$$① \begin{pmatrix} \nu_e & \bar{d}\bar{d}\bar{d} \\ e^- & \end{pmatrix} \qquad ② \begin{pmatrix} u_R\, u_B\, u_G & \bar{u}\,\bar{u}\,\bar{u} & e^+ \\ d_R\, d_B\, d_G & & \end{pmatrix}$$

In the first group we have five fermions; in the second, ten (R = red, B = blue, G = green). The electrical charges of the fermions in each group must add up to 0:

$$① \begin{pmatrix} \nu_e = 0 & \bar{d}\bar{d}\bar{d} = +1 \\ e^- = -1 & \end{pmatrix} = -1 + (+1) = 0$$

$$② \begin{pmatrix} +\tfrac{2}{3} + & \tfrac{2}{3} & +\tfrac{2}{3} \\ -\tfrac{1}{3} + & (-\tfrac{1}{3}) & +(-\tfrac{1}{3}) \end{pmatrix} \qquad -\tfrac{2}{3} + (-\tfrac{2}{3}) + (-\tfrac{2}{3}) + 1$$

$$+\tfrac{3}{3} = 1 \qquad\qquad -\tfrac{6}{3} = -2 \qquad +1$$

$$1 + (-1) = 0$$

Figure 12.7 and Table 12.1 show the relative strength of the forces at different levels of energy and their effective distance for interactions. This simplified presentation of the unification of forces means that at levels of energy of 10^{15} GeV and higher, there is no difference between the strong, weak, and electromagnetic interactions. Once the energy levels dip below 10^{15} GeV, the individual forces come to life.

Table 12.1 The Four Forces of Nature

Unification of force	Force	Relative strength	Effective range (cm)	Acting on	Particles exchanged	Their mass	Their spin
at 10^{19} GeV { at 10^{15} GeV { 100 GeV {	Strong	1	10^{-13}	Quarks	Gluons	?	1
	Electro-magnetic	10^{-4}	Infinite	Electrically charged particles	Photons (γ)	0	1
	Weak	10^{-24}	10^{-15}	Electrons, neutrinos, quarks	Bosons W^+, W^-, Z^0	50–100 GeV	1
	Gravity	10^{-40}	Infinite	All particles	Gravitons	0	2

Unification of the Four Forces

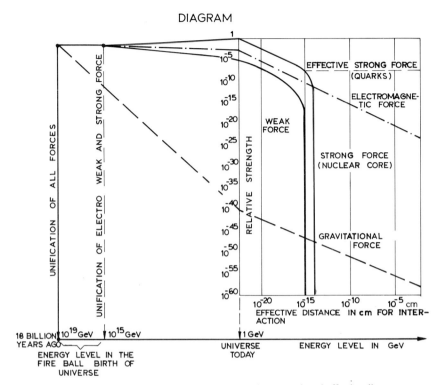

Figure 12.7. The four basic forces of nature: relative strength and effective distance.

It is also assumed that at higher energy levels of 10^{19} GeV when particles in the primordial fireball were very close and temperatures rose to 10^{27} K, the three forces combined with the gravitational constant into one unified force. If we translate the gravitational constant into an energy scale using quantum mechanics, we arrive at the so-called Planck energy, which is 1.1×10^{19} GeV, the same as what we assume to be the minimum level for complete unification of all four forces.

The mathematical background for a complete unification theory is not complete and represents a great challenge to theoretical physicists and cosmologists.

13

Review of Modern Cosmological Theories

13.1. SINGULARITY AND THE BIG BANG MODEL

The generally accepted theory of the birth and evolution of the universe is the classical big bang model. According to this theory, 12–15 billion years ago the universe was born from a cosmic explosion or "big bang" of the "singularity." The singularity contained the entire mass of the universe in the form of elementary particles of matter and electromagnetic energy at infinite temperature, compressed to infinite density in zero space. The mass at infinite density created infinite gravity and, consequently, a complete curvature of space around it. For all practical purposes, the singularity disappeared from space. Nevertheless, in accordance with this theory the entire universe as we observe it today with all the galaxies, stars and planets, white dwarfs and pulsars, quasars and black holes evolved from this gigantic explosion, followed by expansion and cooling down.

As all the physical laws of nature break down at infinite temperatures, compression, gravity, the analysis of the evolution of the universe starts at 10^{-43} sec after the big bang explosion of the singularity. At this time the universe expanded to an approximate size of 2×10^{-33} cm or the so-called Planck length. It takes 10^{-43} sec for a light signal to cross this length. The size of the universe was, at this time, a billion, trillion times smaller than a proton, which has a radius of 10^{-13} cm.

The age of the universe that evolved from the explosion is determined by extrapolating the presently observable expansion velocities of galaxies to time 0, or the time of the explosion.

Astronomical observations indicate that the universe, on a large scale, is homogeneous and isotropic. *Homogeneous* means that the distribution of mass in galaxies or density is uniform throughout the universe. The universe would appear the same to all observers, regardless of their location. *Isotropic* means that it looks the same from all directions, as shown in Figure 13.1.

The discoveries in recent years of enormous local concentrations of mass in the form of clusters of galaxies, called the "great walls," and clusters of quasars put this theory of an isotropic and homogeneous universe in doubt. My own theory of the birth of the universe does not require a homogeneous universe to explain its evolution.

The closed, spherical, and finite universe of the big bang model can be compared to a sphere of an expanding rubber balloon, as shown in Figure 13.2. As the balloon is inflated, the distances between three receding galaxies—A, B, and C forming a triangle—increase but the triangle always maintains the same shape while increasing its size. Although astronomical observations indicate that the universe, on a large scale, is isotropic, we cannot verify whether it is indeed homogeneous as we are unable to travel deep enough into the universe. When regions within a radius of 3 billion light-years away are analyzed, the population of radio galaxies and quasars are equally distributed, within 1%. The standard big bang model relates the fate of the universe to its density. If the actual density were to exceed the critical level (calculated later) of $2 \times 10^{-29}/cm^3$, gravitation which is slowing down the expansion rate would ultimately bring the expansion to a halt. A reverse motion or contraction would follow, resulting in a total collapse of the universe into a cosmic black hole. This would bring to an end the brilliant

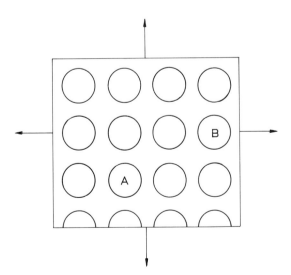

Figure 13.1. Homogeneous and isotropic universe has the same proportions at A and B in every direction.

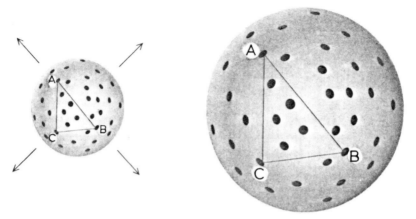

Figure 13.2. Schematic diagram of the balloon "model of expansion of the universe." Each dot represents a galaxy.

birth of the universe, its glorious, though sometimes chaotic development, including all of our knowledge about its operation.

13.2. OBSERVATIONAL PROOF OF THE BIG BANG MODEL

Two predictions proven to be correct by observations have brought about a large acceptance of the big bang theory. The first one is the observed abundance of helium and deuterium in the otherwise hydrogen-filled universe. The second prediction is the much later discovered cosmic background blackbody radiation.

13.2.1. Primordial Helium and Deuterium

Seventy-four percent of matter in the universe consists of hydrogen gas. Helium represents 24% and deuterium (the heavy isotope of hydrogen) represents 1%. Approximately 1% (the heavy elements) was produced much later in the cores of stars. A uniform distribution of helium and deuterium is observed in all galaxies, including our Milky Way, in quasars and nebulas of ionized gas around young stars. This uniform distribution of helium and deuterium indicates that these light elements were synthesized shortly after the big bang explosion and that what we actually observe is primordial helium and deuterium, produced together with hydrogen during the so-called nucleosynthesis approximately 3 min after the big bang.

Helium is also synthesized by fusion of hydrogen in the cores of stars.

However, it is well known that less than 10% of primordial hydrogen has been converted in this way. Also, deuterium is a fragile element and would not survive the high temperatures in the cores of stars. All of this speaks for the theory that the production of these two elements took place in the first minutes after the big bang, when temperatures were still approximately 1 billion K.

13.2.2. Background Radiation

The most dramatic support, however, of the big bang theory was the discovery in 1965 by Arno Penzias and Robert Wilson of a microwave radiation in the millimeter range (Figure 13.3). This radiation, which was predicted much earlier by Gamow, floods the universe and is highly isotropic and uniform in its intensity, better than 1 part in 1000 from any direction measured on earth and in space. This microwave radiation is considered to be the cooled echo of the explosion in the form of primordial blackbody radiation with an energy spectrum

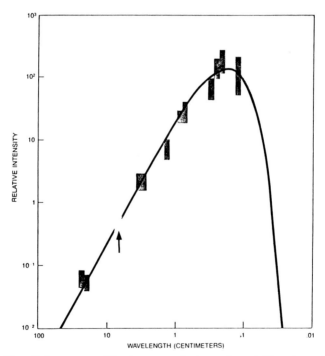

Figure 13.3. The radiation spectrum of the electromagnetic photons decoupled from the particles of matter when the universe cooled down to 3000 K. The spectrum shown was discovered by A. A. Penzias and R. W. Wilson. The radiation now is at 2.9 K. They discovered it at the wavelength marked by the arrow.

at a temperature of 2.7 K. These are the cold remnants of electromagnetic energy from the superhot primeval fireball when there was a perfect equilibrium between matter and radiation and the fireball acted as a blackbody radiation source of heat. A blackbody is an ideal absorber and radiator of radiation of all wavelengths.

The formula for the energy density E of radiation is

$$E_d = aT^4 \quad \text{erg/cm}^3 \tag{13.1}$$

where T is the temperature in degrees Kelvin and the radiation density constant $a = 7.56 \times 10^{-15}$ erg cm^{-3} (degree)^{-4}T. At the present temperature of $T = 3$ K, the energy density is

$$E_d = a(3)^4 \quad \text{or} \quad 6.1 \times 10^{-13} \text{ erg/cm}^3 = 0.38 \text{ eV/cm}^3$$

(1 eV = 1.6×10^{-12} erg.) This is similar to the energy density of starlight coming from the Milky Way. The peak of the relative intensity as shown in Figure 13.3 occurs at a wavelength of 1 mm (0.1 cm).

13.3. THE AGE OF THE UNIVERSE

The big bang theory predicts the beginning of the universe at a definite time in the past, coinciding with the explosion of the singularity. The age of the universe is calculated from the present expansion rate of the galaxies and their distances extrapolated to time $t = 0$.

13.3.1. The Redshift of Light

Although the speed of light remains constant at 300,000 km/sec, regardless of the motion of the emitting star or galaxy, we can establish the modality of motion toward or away from earth by analyzing the spectrum of the incoming light and changes in its wavelength.

If the light source of a galaxy is moving away from earth, the waves are spread over a longer distance and the light recorded has a longer wavelength than the emitted wavelength. This shift of the light to longer wavelength is called the redshift or Doppler effect and the increase in wavelength is directly proportionate to the speed of the observed galaxy. From the amount of the redshift we can determine the velocities at which the galaxies recede from us. However, as it takes considerable time for the light to reach us on earth, we observe the galaxies as they existed at the time the light was emitted. If the distance, for instance, is 1 billion light-years, we see the galaxy 1 billion years back in time. Typical redshift spectrograms are shown in Figure 13.4 indicating the applicable velocities of recession and the shifting of the H and K lines.

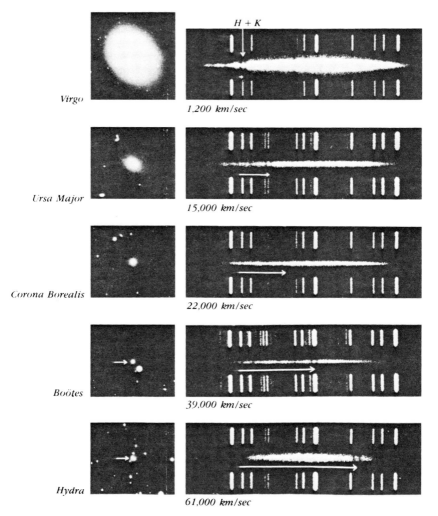

Figure 13.4. The redshift of five galaxies. The spectra of light are the horizontal streaks tapered to the left and right. If the galaxy did not recede, the K and H dark lines of calcium would be the vertical broken line. The amount of redshift or receding speed is shown as the horizontal arrow. (California Institute of Technology)

If the receding velocity of a galaxy is v, then the distance d traversed in time t will be

$$d = vt \tag{13.2}$$

If $t = 1$ sec, $d = v$; in 1 sec the galaxy moved the distance equal to v.

If the incoming wave frequency or the number of waves per second of the emitted light is f, then the time between two successive waves is $1/f$. During that time, the star moved $(1/f) \times v = d$ or v/f centimeters. The observer on earth therefore measures a longer time interval.

The earth-time interval $= f^{-1} + (v/c)f^{-1}$ seconds, where c is the velocity of light. We observe a lower frequency of the incoming light than the emission of the observed galaxy. The ratio of the observed frequency to the actual frequency is:

$$\frac{f'(\text{observed})}{f(\text{emitted})} = \frac{1}{1 + v/c}$$

If we replace the frequency by the wavelength λ, $\lambda = ct$, we obtain

$$\frac{\lambda'(\text{observed})}{\lambda(\text{emitted})} = 1 + \frac{v}{c}$$

or

$$\lambda(\text{observed}) = \lambda(\text{emitted}) + \lambda(\text{emitted})\frac{v}{c} \tag{13.3}$$

$$\frac{v}{c} = \frac{\lambda'(\text{observed}) - \lambda(\text{emitted})}{\lambda(\text{emitted})} \tag{13.4}$$

For instance, if the galaxy Hydra, 1.1 billion light-years away, speeds away from our galaxy at 61,000 km/sec, the wavelength of any spectral line for Hydra is longer than its normal size by:

$$\frac{\lambda'}{\lambda} = 1 + \frac{61{,}000 \text{ km/sec}}{300{,}000 \text{ km/sec}} = 1.2033$$

This mathematical relationship clearly indicates that the recession of an object emitting light results in a longer wavelength or redshift. The relationship between the receding speed v and the speed of light c is defined by Z:

$$Z = \frac{\lambda'(\text{observed}) - \lambda(\text{emitted})}{\lambda(\text{emitted})} = \frac{v}{v} \tag{13.5}$$

When the velocity of the receding galaxy approaches the speed of light, a correction must be made using the Einstein relativistic formula:

$$1 + Z = \frac{1 + v/c}{(1 - v^2/c^2)^{1/2}} \qquad (13.6)$$

If the velocity of recession were to equal the speed of light ($v = c$), $1 + Z = 2/0 = \infty$, the shift Z would be infinite (∞), which means that the source could not be detected. This is the logical and mathematical proof that no object containing matter can move at the speed of light. Only massless electromagnetic radiation can move with the ultimate speed of light.

13.3.2. The Hubble Constant of Recession H_0 as a Basis for Calculating the Age of the Universe

The actual recession speed of stars and galaxies and their distances are determined by using a constant H_0, named after Hubble, which runs from 50 to 100 km/sec per Mpc of distance or 15 to 30 km/sec, for each million light-years distance. Edwin Hubble was the first to demonstrate, based on astronomical observations, that there is a definite relationship between the recession speed of a galaxy and its distance. The larger the distance, the higher is the recession speed. The relationship between recession speed and distance is expressed as

$$v = H_0 d \qquad (13.7)$$

where v is the recession velocity, d is the distance, and H_0 is the Hubble constant.

The age t of the universe can be calculated from the Hubble constant as follows. By Eq. (13.2), the velocity of recession is

$$v = \frac{d}{t}$$

Substituting this result into the Hubble equation $v = H_0 d$ yields

$$\frac{d}{t} = H_0 d$$

$$t = \frac{1}{H_0} \qquad (13.8)$$

The age of the universe, or time t from the big bang explosion, is the reverse of the Hubble constant provided H_0 was constant from the beginning of expansion, which is not the case. Corrections must be made for deceleration. The distances of galaxies are normally measured in parsec (pc). A light-year is the distance traversed by light in 1 year at 300,000 km/sec velocity:

$$1 \text{ light-year} = 0.94605 \times 10^{18} \text{ cm}$$
$$1 \text{ pc} = 3.0856 \times 10^{18} \text{ cm}$$
$$1 \text{ pc} = 3.26 \text{ light-years}$$

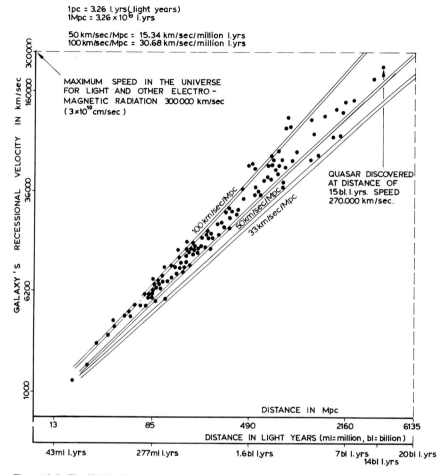

Figure 13.5. The Hubble diagram of expansion of the universe. Data on galaxies and quasars with their distance and recession speed determined from observation.

The Hubble diagram (Figure 13.5) plots the distance and recession velocities of observed galaxies (dots) between the lines of the Hubble constant $H_0 = 33$, 50, and 100 km/sec per Mpc. Most measurements plotted indicate that H_0 is between 50 and 100 km/sec per Mpc. Astronomers believe that H_0 is closer to 50 km/sec per Mpc or 15.33 km/sec per million light-years distance. This means that for each million light-years distance, a galaxy's speed is increasing by 15.33 km/sec. A galaxy at a distance of 10 million light-years would, therefore, have a recession velocity of 153.3 km/sec.

Theoretically, if the most distant galaxy were to move at the speed of light or

300,000 km/sec, which is faster than any object containing matter can move, it would now be at a distance of

$$d = \frac{v}{H_0} = \frac{300,000}{15} \times 10^6 \text{ light-years}$$

(H_0 = 15 km/sec per 10^6 light-years)

$$= 20,000 \times 10^6 \text{ light-years}$$

The universe would thus be 20 billion years old.

Photons and possibly neutrinos and antineutrinos, which move at the speed of light, could theoretically be at that distance. The universe therefore has a maximum theoretical age of 20 billion years. If the Hubble constant were 30 km/sec per million light-years, the universe would be 10 billion years old. As the most distant galaxy, or rather quasar, located in the center of the galaxy was discovered recently at a distance of 15 billion light-years and it would take a minimum of 1 billion years for galaxies to develop and another 2–3 billion for stars to form, the minimum age of the universe based on the most recent discoveries would be 17–18 billion years from the big bang.

The exact age of the universe can be calculated from the rate of expansion, which, in turn, depends on the density of matter. Also, it seems to be obvious that the rate of expansion was much higher at the beginning of expansion when the universe consisted of radiation and elementary particles of matter such as electrons and quarks, and expanded at speeds close to the speed of light. This expansion speed must have been gradually reduced when heavier particles such as baryons and later atoms formed, moving at lower speeds.

As mentioned, the Hubble constant H_0 seems to be between 50 and 100 km/sec per Mpc or 15.33–30.66 km/sec per million light-years distance. The rate of change in the expansion rate caused by gravity can be expressed by a deceleration parameter d_0.

A universe that contains more than the critical mass or critical density is closed and will recollapse; therefore, $d_0 > \frac{1}{2}$. Most astronomers presently assume that H_0 is 50 km/sec per Mpc and d_0 is close to $\frac{1}{2}$.

Different cosmological theories assume different average expansion rates. In the classical big bang theory, using H_0 = 50 km/sec per Mpc or approximately 5×10^{-11} years, the age of the universe or inverse of H_0 as we have already calculated must be less than 20 billion years:

$$\frac{1}{H_0} = \frac{1}{5 \times 10^{-11} \text{ years}} = 20 \text{ billion years}$$

In the inflation theory, which is a modified model of the classical big bang theory and which will be described later, the age of the universe is $t = \frac{2}{3} \times H_0^{-1}$ or approximately 15 billion years. However, the recent discoveries of quasars 14 and 15 billion light-years away put the theory of inflation in doubt.

When Hubble observed galaxies and analyzed their speed of recession, redshift exposures of single galaxies took a long time, sometimes as much as a week. Light from the brightest part of a galaxy was passed to a reflecting grating, recording the spectrum on a photographic plate. A typical galaxy spectrum is really a composite spectrum of millions of stars in the galaxy. Today, in order to investigate the isotropic distribution of redshifts of thousands of galaxies, new methods are used.

In Figure 13.6, over 250 galaxies can be seen. It is the central region of a very rich cluster called Coma, which may contain over 10,000 galaxies.

The increased efficiency in redshift analysis is due to two innovations. The first is the image intensifier, which converts the incoming photon image into an equivalent electron image. The electrons are then highly accelerated to induce energy and fall on phosphor, where the image is converted again into much intensified light. The brightness and resulting spectra may be several thousand times clearer.

The second innovation is an instrument that uses semiconductors. A large number of capacitors accumulate electrical charge produced by the incoming light. The composite charge is then fed directly into a computer for analysis. Results are 100 times better than the most sensitive photographic plates. Redshift results can also be obtained with radio telescopes receiving the 21-cm hydrogen wave.

13.3.3. The Observable Data

Large telescopes and better observational methods have enabled researchers to penetrate deeper than ever (Figure 13.7). During 1986–1988, many quasars were detected at distances of more than 12 billion light-years. In the more than 12-billion-year journey to earth, the light from these very old quasars extended its wavelength over 400% due to the fast recession speed. Z, the value of the expansion of the universe during this time, is obtained by dividing the percentage shift by 100: $Z = 400/100 = 4$. The larger the Z value, the more distant the observed object is. The two most distant quasars have a Z value of 4.40 and 4.43. Until a few years ago, the most distant galaxies were observed at distances of 4–6 billion light-years visually, and at distances of 10 billion light-years with radio telescopes.

Recently, through improved observation, primeval galaxies have been discovered at 14–15 billion light-years away and a globular star cluster named NDC-6752 at 14 billion light-years. At the time of writing, a quasar 15 billion light-years away was photographed (Figure 13.8). With the Hubble recession parameter of $H_0 = 15.33$ km/sec per million light-years, the cluster recedes at the

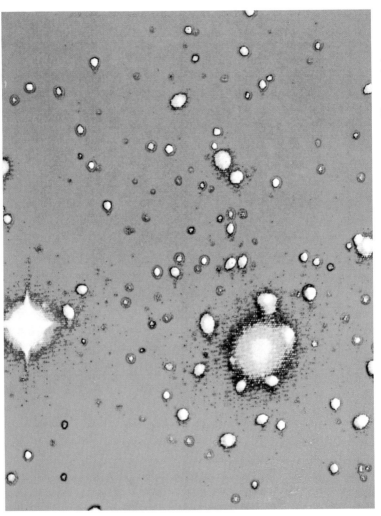

Figure 13.6. A large cluster of galaxies Coma. (National Optical Astronomy Observatories)

Review of Modern Cosmological Theories

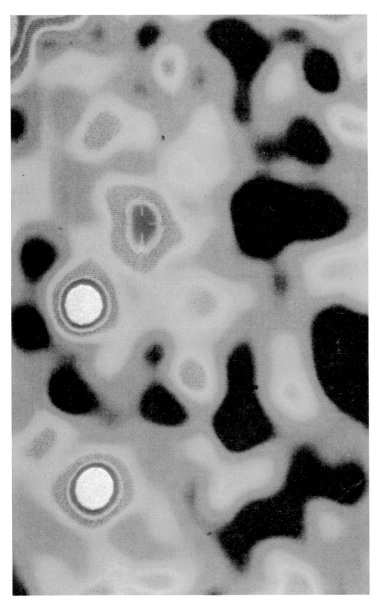

Figure 13.7. False-color infrared image of high redshift galaxy discovered by deep infrared imaging. (R. Elston, University of Arizona)

enormous velocity of 270,000 km/sec or 90% of the speed of light. If we estimate that the first galaxies started to form 1 billion years after the big bang explosion and the first stars formed 3–4 billion years later and started emitting light, we have at best detected with telescopes galaxies and quasars 3–5 billion years from the date of creation. Based on these discoveries, the age of the universe of 17–18 billion years seems to be a reasonable assumption.

More accurate observations were scheduled with the Hubble space telescope

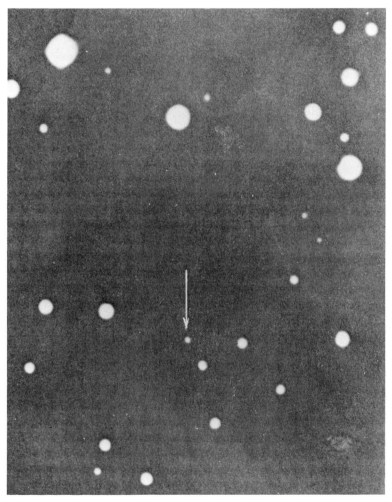

Figure 13.8. Quasar OH471 with receeding speed of $0.9c$ (270,000 km/sec), 15 billion light-years away. (California Institute of Technology)

launched in 1990 without interference of the earth's atmosphere, which would enable us to see perhaps the first stars created in the universe. However, at the time of editing this book the Hubble's main mirror was found to be defective and it will take considerable time, if at all, to correct the deficiency to the great disappointment of astronomers.

13.4. THE FRIEDMANN MATHEMATICAL THEORY OF THE BIG BANG AND SINGULARITY

It was the Soviet scientist Aleksandr Friedmann who developed in the 1920s the mathematical background for the classical big bang theory, starting with the singularity. If we imagine part of the cosmic space as a sphere of galaxies of radius R (see Figure 13.9) that is expanding with the universe after the big bang and make an assumption that the total mass M of the universe inside the shell remains constant with more or less uniform density d, equal to the mean cosmological mass density, then the *total energy of all of the matter must remain constant*. In this discussion, we neglect the pressure of the matter.

Using the law of energy conservation, the sum of the kinetic energy E_K of the expanding universe plus the gravitational potential energy E_G must be constant at all times. The kinetic energy of the expanding mass M with velocity v is

$$E_K = \tfrac{1}{2}Mv^2 \tag{13.9}$$

For a mass equal to unity, the kinetic energy is $E_{K1} = \tfrac{1}{2}v^2$. The direction of the kinetic energy coincides with the direction of the expansion of the universe.

The gravitational potential energy E_G of the shell acting in opposite directions to the kinetic energy can be expressed as

$$E_{G1} = -G\frac{M}{R} \tag{13.10}$$

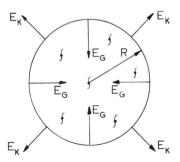

Figure 13.9. Sphere of galaxies of radius R. E_G is the gravitational potential energy. E_K is the kinetic energy of the expanding universe; arrows show the magnitude.

where G is the new Newtonian gravitational constant and R is the radius of the sphere. The gravitational energy is considered to be negative as the universe will come to a halt and reverse its direction if the mass of the universe is more than critical. The gravitational energy acquired during the fall toward the center will be largest at the center where the kinetic energy will be reduced to 0. In the case of a singularity, therefore, with a radius $R = 0$, the gravitational energy E_{G0} becomes infinite. This is the mathematical background using simple equations to provide the "logic" for the infinite gravity at $R = 0$. At $R = 0$

$$E_{G0} = G\frac{M}{0} = \infty = \text{singularity}$$

We can now write the basic mathematical equation that governs the Friedmann big bang model, based on Einstein's general relativity:

energy of expansion + potential gravitational energy = constant

$$\tfrac{1}{2}v^2 + -G\frac{M}{R} = \text{constant}$$

The sum of the two types of energy must be constant with time. The constant, which is also called the curvature constant K, is actually the expression of the total average energy held by 1 gram of material in the universe. The constant K can be 1, -1, or 0, depending on how much matter is in the universe. The fate of the universe depends on K, which indicates how much mass is in the universe.

The three alternative scenarios are showin in Figure 13.10. The universe is closed when its mass is larger than critical and $K = +1$, it is flat when the mass is critical and $K = 0$, and it is open, expanding forever when the mass is less than critical and $K = -1$.

Using K we can write the Friedmann equation as:

$$\tfrac{1}{2}v^2 - G\frac{M}{R} = -\tfrac{1}{2}K$$

If we now apply the Hubble equation for the expansion velocity of galaxies

$$v = H_0 R$$

and use for mass M = volume × density (d)

$$M = \tfrac{4}{3}\pi R^3 d$$

we get

$$\tfrac{1}{2}H_0^2 R^2 - G\tfrac{4}{3}\pi R^2 d = -\tfrac{1}{2}K$$

Dividing by R^2 we obtain

$$\tfrac{1}{2}H^2 - \tfrac{4}{3}\pi G d = \frac{K}{2R^2}$$

Review of Modern Cosmological Theories

Multiplying by 2 we obtain

$$H^2 - \tfrac{8}{3}\pi G d = -KR^{-2} \tag{13.11}$$

Let us consider the three models of the universe:

flat–open (Einstein–de Sitter) $K = 0$
open–infinite (Friedmann–Lemaître) $K = -1$
closed–cycling (Friedmann–Lemaître) $K = +1$

Alternative 1: $K = 0$, universe is flat–open and density is exactly critical:

$$H_0^2 - \tfrac{8}{3}\pi G d = 0 \quad (K = 0, \;\; -KR^2 = 0) \tag{13.12}$$

$$d_{crit} = \frac{3H_0^2}{8\pi G} \tag{13.13}$$

We can calculate the critical density of the universe d_{crit} with $H_0 = 15 \times 10^5/10^{24}$ cm, $G = 6.67 \times 10^{-8}$ cm^3 g^{-1} sec^{-2}, and 1 light-year = 9.5×10^{17} cm (1 million light-years ~ 10^{24} cm):

$$d_{crit} = \frac{3}{8\pi}\left(\frac{15 \times 10^5}{10^{24}}\right)^2 \frac{1}{6.67 \times 10^{-8}} \cong 4.5 \times 10^{-30} \text{ g/cm}^3$$

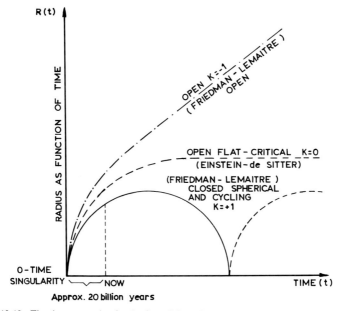

Figure 13.10. The three scenarios for the fate of the universe, depending on the amount of mass.

As the weight of a hydrogen atom is 1.66×10^{-24} g, the critical density in terms of particles is $d_h = 2.7 \times 10^{-6}/\text{cm}^3$ or, on the average, 2.7×10^{-6} hydrogen atoms per cm^3 of space or 2.7 hydrogen atoms in 1 million cm^3 of space.

Alternative 2: $K = +1$, *universe is closed and mass is larger than critical*:

$$H_0^2 - \tfrac{8}{3}\pi Gd = -R^2 \qquad (K = 1, \quad -KR^2 = -R^2)$$
$$H_0^2 = \tfrac{8}{3}\pi Gd - R^2 \tag{13.14}$$

It is clear that if the universe expands to its maximum (R^2 max), it will finally reach a level equal to $\tfrac{8}{3}\pi Gd$. The rate of expansion of H_0 will become 0 and there will be no expansion. The universe will come to a standstill and then recollapse into a cosmic black hole or singularity, according to the standard cosmological theory. As we will see later, my theory advances a more logical outcome and eliminates the singularity as the birthplace of the universe.

Alternative 3: $K = -1$, *universe is open and mass is less than critical*:

$$H_0^2 - \tfrac{8}{3}\pi Gd = +R^2 \qquad (R = -1, \quad -KR^2 = +R^2)$$
$$H_0^2 = \tfrac{8}{3}\pi Gd + R^2 \tag{13.15}$$

In this case with the increasing radius of the expanding universe, H will never become 0 and the universe will expand forever with decreasing density d.

13.5. THE FOUR BIG QUESTIONS ABOUT THE BIG BANG THEORY

After many years of intensive analysis of the classical theory of the big bang, four basic questions remain mysteries:

1. Where did the enormous energy required to create all of the particles of matter contained in the universe originate?
2. How did all matter and radiation contained in the universe get into the singularity in the first place? So far, there are no clues or even speculative views as to the origin of matter and radiation in the singularity. Several attempts were made in proposing theories for the creation of matter, which we will review later, but they all seem to fall into the categories of metaphysics or even science fiction.
3. How could the gigantic amounts of matter and radiation be compressed to *infinite density* in *zero space*—a condition where all known physical laws of nature break down? It is my opinion that any cosmological theory, even

if backed up by mathematical logic that include infinities, cannot be considered a universally applicable theory of creation.
4. How could the universe have started isotropically and uniformly everywhere?

13.6. SINGULARITY AS A COLLAPSED "FORMER" UNIVERSE

One theory suggests that the "singularity" was in essence a cosmic primordial black hole created as a result of a gravitational collapse of a universe that existed before, expanded, and finally contracted in a reverse cycle. This solution of a cycling universe does not, however, explain how matter was created the first time around. In addition, according to the latest theories, black holes with large masses do not explode.

The idea of the singularity was conceived by extrapolating observational data of the expanding universe to the time zero, using the Einstein theory of relativity as described in the Friedmann mathematical analysis. Einstein himself, however, disliked the idea of a singularity and expressed serious doubts, in the fifth edition of his book *The Meaning of Relativity* in 1951, as to the validity of his theory for the special conditions of extreme levels of density, heat, and field energy that prevailed during the time of creation. Einstein's first doubts are expressed on p. 118 in Appendix 1 of the 1951 edition, where he writes ". . . For every state of non-vanishing ("spatial") curvature there exists, as in the case of vanishing curvature, an initial stage where radius "G" = 0, where expansion starts. Hence, this is a section at which the density is infinite and the field is singular. *The introduction of such a new singularity seems problematical in itself.*" ". . . It may be plausible that the theory is for this reason inadequate for very high density of matter. It may well be the case that *for a unified theory there would arise no singularity.*" Einstein continues on p. 123 in Appendix 1 ". . . One may not therefore assume the validity of the equations for very high density of field and of matter, and one may not conclude that the beginning of the expansion must mean a singularity in the mathematical sense. All we have to realise is that the equations may not be continued over such regions. . . ."

13.7. SINGULARITY AS A BLACK HOLE OF ELECTROMAGNETIC ENERGY

Recently, a new strange version of the singularity big bang model was presented by Gary Bennet in *Astronomy* (August 1988). According to his interpretation, the singularity was an infinitely dense and hot concentration of electromagnetic energy that erupted in a titanic explosion, creating all space, time, matter, and electromagnetic radiation. To review this interesting idea we must analyze the relationship between radiation density and prevailing temperature. Density of radiation depends on temperature:

$$d_r = \frac{aT^4}{c^2} \tag{13.16}$$

$$T = 4\left(\frac{d_r c}{a}\right)^{1/2} \tag{13.17}$$

where a is the radiation density constant [= 7.56×10^{-15} erg cm^{-3} (degree)$^{-4}$. In accordance with this equation, if the density is infinite ($d_r = \infty$) the temperature also is infinite ($T = \infty$), which does not make any sense.

Radiation can be trapped together with matter when stars collapse into black holes. This results in enormous gravitational forces and curved space, preventing radiation from escaping. This theory seems to me to be another metaphysical exercise in science fiction, unless the Creator decided to create the universe in such an unusual way, not complying with its own laws of nature. I have my doubts that this was so. In any case there is no explanation as to how radiation came into the black hole in the first place.

13.8. GOD AS CREATOR

The difficulty with all cosmological theories, including the classical big bang model, is the unresolved question, where did the enormous energy required for particle creation come from? My own cosmological theory presented later in this book gives a solution to this and other questions and eliminates the troublesome singularity. The primordial fireball had finite dimensions and exploded before the mass of particles and radiation reached a point of no return and irreversibly collapsed into a black hole.

The Catholic Church adopted in 1951 the classical big bang model, declaring the creation of the singularity as an Act of God. An interesting meeting, arranged by the Jesuits, was held in Vatican City in 1981 to mark the 30th anniversary of the adoption of the theory. His Holiness, Pope John Paul, took part and discussed the Act of Creation and all other related theories with the world's most renowned cosmologists and astrophysicists. During the meeting the Pope suggested to the visiting scientists to direct their efforts toward trying to unravel the events that followed the big bang explosion when matter and radiation seemed to behave according to known physical laws of nature. He took the position that the "singularity" was created by God, and therefore details will remain forever beyond the comprehension of the human mind. He also recommended that this Act of God should not be investigated.

The scientists concluded that the never-ending quest to explain the inexplicable will continue until the puzzle of creation is resolved and a final theory developed that ultimately is backed up by observational proof and simulated experimental analyses.

There are some cosmologists who believe that because it is so difficult to comprehend or explain the singularity, it is possible that God could have started the universe in such an unusual way.

The history of science indicates that all major discoveries reflect an underlying order that seems to be inspired by a Divine Being. Why, therefore, would God choose such an incomprehensible way for the creation act and then let the universe evolve in accordance with defined physical laws? It would seem more logical to assume that God, having a choice, would have selected the beginning as well as the era before and after creation to be governed by the same underlying laws and rules that apply to the universe today.

13.9. CONTRADICTIONS IN THE BIG BANG THEORY WITH AN EXPLODING SINGULARITY

In addition to the mystery of the singularity discussed in a previous chapter, the big bang theory does not give answers to many other questions and some solutions are in conflict or do not conform with astronomical observations. Here we review the difficulties and the contradictions with recognized and experimentally proven physical laws of nature.

1. *Singularity as a primordial cosmic black hole*. Based on indirect observations and mathematical calculations, *large black holes do not explode* and actually have been determined to be stable. This means that a black hole, characterized by mass and angular momentum, is a permanent final state. The big bang theory does not explain the cause of explosion of the singularity.

2. *Infinite curvature of the singularity would isolate the cosmic black hole from the cosmos*. The enormous concentration of matter resulting in an infinite gravitational field would have curved the space around the singularity to a point where the singularity would disappear from the cosmos altogether.

3. *The big bang theory claims that there was no space, as such, before the big bang explosion, that time was zero, and that everything started with the big bang*. The parameters of the singularity and the present big bang theory insist that there was no space, as such, and no time before the explosion which took place in zero space. As time started only at the big bang, it is impossible to look beyond.

Space was created simultaneously with the expansion of the huge mass. This is another highly contradictory statement made by some of the world's leading scientists who claim that space around the singularity was infinitely curved and perhaps this, alone, caused particles to be created around the singularity 10^{-43} sec after the explosion. It seems to me that if space did not exist prior to the explosion and expansion of the singularity, it could not be infinitely curved. The singularity is often described as an infinitely compressed mass in zero space. Also, it is well known today that, as a direct result of Heisenberg's relativistic quantum physics, the so-called empty space is filled with virtual particles which can become real particles if through an outside source a high-energy field of electromagnetic energy is introduced, at least equivalent to the rest mass of the particles. In addition, large topographical fluctuations occur in space at Planck distances of 10^{-33} cm. Thus, it seems logical to assume that just as time existed before the big bang, space also existed in the form of what we call the vacuum of space.

4. *Einstein's theory of relativity and all known physical laws break down.* Einstein himself, as we cited from his book, doubted that his theory could be extended to the time of creation when extreme temperatures and densities of matter and radiation prevailed. He tried to eliminate the singularity for the last 30 years of his life, as many other specialists and mathematicians tried after him, to amend the theory of relativity and make it universally applicable, using proven physical laws of nature. I do not think that God played tricks with nature during the creation of the universe.

5. *How did matter get into the singularity in the first place?* There is no explanation for how matter got into the singularity and what happened before time 0 when matter had to be created somehow and collapsed into a singularity or primordial cosmic black hole.

6. *Where did the energy come from to generate the particles of matter that collapsed into the singularity?* The energy of the universe, if inherited from the fireball, is extremely powerful. It is clear, therefore, that the energy that was transformed into particles of matter must have been equally powerful. It could not have come from nothing, as this would violate the basic law of nature, the law of preservation of energy.

7. *The big bang theory does not explain why the universe is isotropic and homogeneous in all directions.* The observable universe is homogeneous and isotropic, as explained earlier. There are two possibilities to explain this uniformity. The universe either started with complete isotropy in a smooth and simple way, or it started chaotically and later on in the early expansion became isotropic. Based on the Hubble theory of expansion, proven by observations, it can simply be calculated that in the early universe, after the big bang, a large portion of the particles were not in contact with each other and therefore it is difficult to imagine how any irregularities in the total mass could have been smoothed out.

This problem is called "the horizon problem" can be explained as follows. According to the standard big bang theory, there would actually be two sizes to the universe. The first size is calculated from the time of expansion t or the age of the universe and the velocity of expansion, which, in the first few seconds, was close to the speed of light c or $A_R = ct$. This is shown schematically in Figure 13.11. Based on this formula the A_R radius of the universe, 1 sec after the explosion, was

$$A_R = 300,000 \frac{km}{sec} \times 1 \text{ sec} = 300,000 \text{ km}$$

This in itself was an enormous expansion in the size of the universe, considering that at 10^{-45} sec after the explosion the entire universe had the size of the Planck distance of 10^{-33} cm.

As A_R 300,000 km = 3×10^{10} cm, the universe expanded in 1 sec

$$\frac{3 \times 10^{10}}{10^{-33}} = 3 \times 10^{43}\text{-fold}$$

which is difficult to imagine and even more difficult to comprehend. But this is not enough!

Using the Hubble formula for the velocity of expansion $v = H_0 R$, the radius R 1 sec after expansion can be calculated using $H_0 = 15.34$ km/sec per million light-years, 1 light-year = 0.946×10^{18} cm = 0.946×10^{13} km, and velocity of the universe 1 sec after the big bang $\sim$ the speed of light, i.e., $v = c = 300,000$ km/sec. Based on the Hubble equation $v = c = H_0 R_H$,

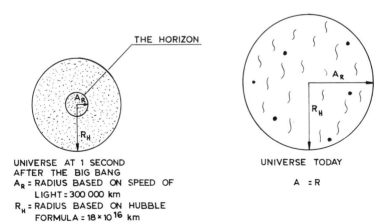

Figure 13.11. The horizon problem just after the big bang.

$$R_H = \frac{c}{H_0} = \frac{300{,}000 \text{ km/sec}}{(15.34 \text{ km/sec})/0.946 \times 10^{13} \text{ km} \times 10^6}$$

$$= \frac{300{,}000 \times 0.946 \times 10^{19} \text{ km}}{15.34}$$

$$= 18 \times 10^{16} \text{ km}$$

The expansion on the scale of R_H was 10^{16} times larger than the radius A_R based on the distance traversed by the light in 1 sec.

Based on the Hubble formula, the universe expanded in less than 1 sec from 10^{-33} cm to 18×10^{16} km or $(18 \times 10^{16})/10^{-33} = 18 \times 10^{44}$. One could also say that the universe expanded infinitely larger from the singularity, which had 0 radius.

The only way this can be explained is to assume that the explosion took place everywhere simultaneously—not like explosions on earth, from the center. In other words, every particle sped away from every other particle everywhere, at a speed close to that of light. I must stress that this is the classical big bang theory and not my own theory of creation, which is described in Chapters 15–25.

Because no signal can travel faster than the speed of light, no information can be obtained from the large area beyond the radius A_R called the horizon. Consequently, a large portion of particles in the early universe were outside the horizon, could not interact, and could not smooth out irregularities created during the explosion.

As the universe expanded, A_R increased faster than the radius R_H of the universe, which was slowing down due to gravity when heavier nuclear particles and, later, atoms of hydrogen and helium were created.

Today, $A_R = R_H$. If, however, the universe started and expanded isotropically, the "horizon problem" explained here remains an interesting but irrelevant analysis of mathematical formulas applying to the big bang theory.

13.10. AMENDED BIG BANG THEORIES

There are several other cosmological theories that try to amend the classical big bang model but recognize the singularity. These theories try to eliminate one or more difficulties and inconsistencies with the model but they all deal with events after the explosion and do not shed any light on the controversial singularity. There are others who work hard on the mathematical portion of the theory using Einstein's field equations, extrapolating them to time zero of creation and trying unsuccessfully to eliminate the mathematical singularity. Those who succeed in this complicated undertaking and amended the Einstein formulas found that, under these circumstances, the whole theory fell apart.

13.10.1. The Inflation Theory

One of the well-known cosmological theories is called inflation, a basic theory that claims that the universe underwent a very short period of enormous expansion at 10^{-35} sec after the big bang, even larger than described earlier in this chapter.

Alan Guth, in 1980, introduced into classical big bang theory a new idea dealing with the behavior of the expanding mass shortly after the explosion of an object that replaced the singularity but was never defined. It needs to be resolved by quantum gravity. According to his theory, as shown schematically in Figure 13.12, the radius of the universe, shortly after the explosion, was much smaller than in the standard model. At 10^{-35} sec, however, the expanding mass of matter and radiation suddenly underwent rapid expansion, called inflation, which ended at 10^{-30} sec. After the short inflationary period, the radii in both theories were the same.

The Guth theory claims that when the singularity exploded, the universe expanded, particles and radiation were cooled down, and the energy associated with the vacuum of space-time was also "cooling" down.

In the standard singularity model, there is no vacuum of space around the singularity; no space at all. All space is being created after the explosion,

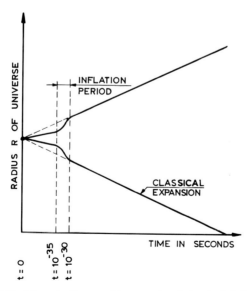

Figure 13.12. Schematic diagram of the expansion during the inflation period.

simultaneously with the expansion of the universe. The inflation theory introduces space existing around the singularity. The theory claims that at approximately 10^{-35} sec after the explosion, the energy of the vacuum of space, which normally also cools down, suddenly remained at the same "supercooled" level, allowing an unusually rapid expansion with a magnitude of 10^{30} for 10^{-5} sec.

In other words, the vacuum of high temperature caused by the high heat of the just-created universe became a "false" vacuum whereby the density suddenly remained constant, in spite of the expansion, and more energy was created to fill the empty space and keep the energy density constant. The false vacuum created negative work, causing energy to flow in.

This false vacuum, which created itself suddenly at 10^{-35} sec after explosion, caused the runaway expansion of the universe or inflation. After this short period, inflation stopped, the vacuum energy rapidly decayed to the original rate of "cooldown," and the expansion of the universe continued at the levels of the standard model. If the universe started in an irregular nonisotropic way, during the inflation period all particles were in contact as the radius R_H of the universe was smaller than the horizon radius A_R (shown in Figure 13.11 without inflation) and, in this way, any irregularities could be smoothed out during this inflation period.

Linde, Albrecht, and Steinhardt amended the Guth theory by introducing the idea of many individual bubbles of space being blown up during the inflation, one of which is our universe. This theory also tries to resolve the problem of the missing magnetic monopoles. These particle-magnets, with a single north or south pole, have been predicted to exist based on the unified theory of electromagnetic, weak, and strong nuclear forces. Magnetic monopoles, however, have not been discovered in spite of intensive efforts and the construction of special detectors.

The amended inflation theory explains that the high density of monopoles in the universe, just after the explosion, became diluted in the rapidly expanding bubbles of space and, therefore, cannot be found. The theory also predicts large fluctuations in the density of matter after the period of inflation, which contradict the uniformity of the background cosmic radiation and which are also much too large for creation of galaxies, as observed. If matter in the universe after inflation had been as irregular in density as predicted by the new inflation theory, the irregularities would have grown later to such an extent that they would have turned into massive black holes and very few galaxies would have been formed.

It is my opinion that for galaxies to be formed, the universe must have started with near-isotropic matter distribution, except for some local density fluctuations. These denser areas with higher gravity levels attracted just enough additional mass to collapse into protogalaxies, evolving into galaxies and stars.

13.10.2. What Triggered Inflation?

A further analysis of inflation theory by Guth and others raised the question of what triggered the inflation process. A strange explanation was devised introducing so-called condensation energy theory. The story goes as follows:

A phase transition, similar to the transition of vapor into liquid which releases energy called condensation energy, took place when the singularity exploded and the hot universe expanded and cooled. The theory goes further by saying that the enormously energetic singularity released, with the explosion and phase transition, a powerful condensation energy which forced the universe to inflate. After the short inflation period was over, the remaining condensation energy transformed itself into matter, based on the Einstein equivalent equation of $E = mc^2$ or $m = E/c^2$.

According to this analysis, the singularity contained only a small quantity of matter, maybe 20 kg, but a gigantic amount of energy, which was released as condensation energy and which created after the inflation what we call the matter contained in the universe.

This new interpretation was described in the October 1989 issue of *Astronomy* by George Greenstein. As a logical consequence of these thoughts, Greenstein suggests that each black hole in the universe will explode sooner or later in a similar way to the singularity and trigger the creation of a new universe by transforming the black hole's condensation energy, released during the explosion and subsequent phase transition, into enormous amounts of matter.

A comparatively small amount of matter contained in a black hole, created from the core of a dead star (20–49% of total content), can act as a seed and trigger the release of enormous condensation energy to create a new universe.

The Einstein equivalent theory expressed in the $E = mc^2$ equation, qualifies that matter is just another form of energy but does not allow violation of the energy conservation law. The total energy contained in a dying star, from which only a partial black hole is created, cannot under any circumstances be transformed or trigger the creation of an entire universe with billions of galaxies and trillions upon trillions of stars.

13.11. OTHER COSMOLOGICAL THEORIES

13.11.1. The Free Lunch Theory of Creation

Contrary to the classical singularity theory where vacuum of space did not exist and was created simultaneously with the expansion of the universe after the big bang, this theory, promoted by Y. Zeldovich and A. Starobinsky in the USSR and E. Tryon in the USA, claims that the universe was created from a vacuum fluctuation. The normal appearance and disappearance of virtual particles, which become real particles in the presence of an outside source of energy, all described in detail in Chapter 14, is the basis of this theory of creation. The theory claims that all of the particles contained in the universe were created suddenly and coincidentally as a long-lived vacuum fluctuation. The explanation for the long period of sustaining the existence of the matter and the universe, in comparison to the short-lived appearance of virtual particles, is justified by the following argument.

The positive, kinetic energy in all of the expanding matter of the universe is fully counterbalanced by the opposite-acting gravitational potential energy, resulting in the total energy of the universe being equal to 0. It is for this reason that the lifetime of the accidental vacuum fluctuation may be infinite.

A variation of this theory has been prepared by R. Brant, P. Englert, E. Gunzig, and P. Spindel of the University of Brussels. In this version, an accidental fluctuation in the vacuum of four-dimensional space-time caused the appearance of a single but extremely heavy particle–antiparticle pair with a mass of 10^{19} GeV. The appearance of this heavy superparticle triggered a self-propagating process of creation of more and more similar superparticles, which created a fireball. Space became heavily curved around the fireball by gravitational forces, which were overcome by internal forces, causing the fireball to explode. The superparticles of 10^{19} GeV mass (10^{19} times heavier than protons of 1 GeV mass) later on decayed into quarks, electrons, and photons, which eventually combined into protons, neutrons, and finally atoms.

The universe was born out of a vacuum fluctuation. It is acknowledged experimentally, in a phenomenon called polarization of the vacuum, that empty space or pure vacuum is filled with virtual (to be) particles, which appear spontaneously as a result of fluctuation of the vacuum and disappear. Particles created in this way such as an electron–positron pair are annihilated almost as soon as they appear and their presence cannot be detected directly. The larger their energy or mass, the shorter their existence is. The uncertainty principle of Heisenberg's quantum theory requires that the virtual particles disappear into the vacuum. The phenomenon can be explained in such a way that, in effect, a virtual particle borrows a quantity of energy but must repay it before the shortage can be detected. Vacuum can therefore be described as space with no real particles in it. It has been proven, however, in high-energy-particle collisions that virtual particles become real particles if they are confronted with an outside source of energy, equivalent to at least the rest mass of the particles. The Zeldovich–Starobinsky–Tryon theory claiming that the universe emerged from vacuum fluctuations does not identify the source of energy required to sustain the emerging virtual particles from the vacuum. They simply say that this particular vacuum fluctuation somehow ran away with itself and created all of the particles out of empty space.

The mathematical justification for the free lunch theory of creation calls for a closed universe with critical positive mass energy, fully balanced by the negative gravitational energy pulling the matter together. Consequently, the positive mass energy ($mc^2 = E$) is being canceled out by the equivalent negative gravitational energy, resulting in the total energy of the universe being zero with no apparent violation of the law of conservation of energy.

Using the quantum, Heisenberg uncertainty relationship between energy and time

$$\Delta E \times \Delta t = \hbar \quad \text{(Planck's constant)}$$

a zero energy value would result in an infinite lifetime duration of such a universe:

$$\Delta t = \frac{\hbar}{\Delta E} = \frac{\hbar}{0} = \infty$$

This, however, is in contradiction to the basic principle of the theory, which calls for a closed universe. Such a universe, born from the vacuum fluctuation, would collapse into a singularity, explode, expand, and recollapse into a cosmic black hole and disappear from the cosmos. The time for this process would, of course, be finite and therefore contradicts the quantum theory, which calls for an infinite life span of a universe with zero energy.

The second notable flaw of the theory is the absence of the enormously powerful creation energy which would be required to sustain the virtual particles emerging from the empty vacuum of space and turn them into all of the known numbers of particles in the universe.

Nevertheless, in spite of its science-fiction characteristics, it is the first theory that attempts to explain how the particles of the universe got into the singularity in the first place.

When the Guth theory of inflation became known, Tryon's "free lunch" theory was amended, mainly to reduce the serious doubts in the possibility of all of the particles of the universe being born simultaneously from the vacuum fluctuation event.

All that would be required now is to create a limited number of particles contained in an extremely small space, smaller than a proton (10^{-13} cm), that would somehow explode. In approximately 10^{-35} sec after the big bang, it would then be blown up in an enormous flash to a gigantic size, as described in the inflation theory, while the large Higgs energy field would turn the vacuum into large "X" particles of matter that later on would decay into known elementary particles such as quarks, electrons, neutrinos, and photons—the basic constituents of the present universe. The explanation of the Higgs energy of vacuum is given later.

Tryon described his theory in 1973 in *Nature*, calling it "the simplest and most appealing cosmological theory."

13.11.2. Another Variation of the Free Lunch Theory of Creation

E. Gunzig, J. Geheniau, and I. Prigogine present a theory where the universe emerges, not from a singularity, but from an instability of the vacuum of space that leads to the creation of matter. The matter in turn collapses into small black holes of approximately 50 times the Planck mass:

$$M = \left(\frac{\hbar c}{G}\right)^{1/2} = 2.2 \times 10^{-5} \text{ g}$$

$$50M = 110 \times 10^{-5} \text{ g}$$

The energy required to create the black holes is extracted from the negative vacuum. These black holes evaporate during an inflation phase. The energy transfer from gravitational curvature of space caused by the black holes creates matter and this is also called the "free lunch" theory.

There are basically two transition phases in this cosmological theory. The first is caused by an instability of the vacuum of space, the second is the standard transition from an inflationary state to the expansion phase, matter–energy model.

There is no observational or experimental proof for such a theory; it is an exercise in a mathematically backed up puzzle, an effort directed mainly toward the elimination of the malaise of singularity.

13.12. THE THEORY OF SUPERSTRINGS OR THE THEORY OF EVERYTHING (TOE)?

On and off for over 30 years, the supporters of the string theory, which considers pointlike particles of matter such as quarks or electrons as excitations of strings, made numerous claims of how the theory can eliminate all anomalies and infinities of Einstein's general theory of relativity and quantum mechanics, indicating that they could unify both theories and the four forces of nature as well as explain the origin of matter and the big bang birth of the universe.

Einstein's general theory of relativity has related gravity to the space structure on a large scale and enabled the evolution of our universe to be described. Quantum mechanics on the other hand deals with the microspace of elementary particles, as small as 10^{-17} cm (electron) and beyond, at Planck's microdimensions of 10^{-33} cm. At these distances it is impossible to integrate Einstein's general theory as gravity becomes infinite. Also, the Einstein concept of space-time consisting of infinite numbers of points, regardless of whether the distance between two points is 10^{-33} cm or 10^{64} cm, would have to be abandoned.

As to the unification attempts of the four forces of nature, quantum mechanics was able to unify the weak, strong, and electromagnetic forces but failed to find a resolution to include gravitation.

After many years of ups and downs from the various string theories, which all regard pointlike elementary particles of matter as different types of excitations of strings, a new variation of the theory was formulated in 1984 by John Schwartz, Michael B. Green, and Joel Shferk. Called superstrings, this theory claims to be capable of unifying all four forces of nature and eliminating all anomalies and infinities of Einstein's general theory of relativity.

13.12.1. Strings

A single string is extremely tiny and one-dimensional, approximately 10^{-32} cm across or 10 million, trillion times smaller than a proton (10^{-13} cm). It can vibrate similarly to a violin string. Each point along the string vibrates and the string is made up from an infinite number of points. The modes of vibration depend on the tension of the string. Each vibration mode corresponds to a given particle. Energy or mass of the string is determined by the vibrational mode. According to this theory, the various elementary particles such as electrons, quarks, neutrinos and nuclear particles such as protons, neutrons, and many others are different modes of the same strings.

The same applies to bosons responsible for mediating interactions of particles such as photons, gluons, W^+, W^-, and Z^0 particles, and gravitons mediating the electromagnetic, strong, weak nuclear force, and gravitational variations of vibrating strings.

There are basically two types of strings, open and closed, both approximately 10^{-32} cm and one-dimensional, as shown in Figure 13.13. Open strings have charged end points such as an electrical charge. Particles associated with open strings vibrating to different modes include massless spin-1 gauge particles such as photons, but not the graviton. Vibrational modes of closed strings include the electron and also the graviton—a massless spin-2 particle—so far not discovered, responsible for gravitational interaction. The vibrational frequencies of superstrings depend on the tension of the string. The tension is measured in energy per unit length or mass, squared (m^2) in basic units. As the superstring theory claims to unite all four forces of nature including gravity, the basic string tension must be related to the Planck energy which is the only dimensional parameter for gravity at such small distances as the string (10^{-32} cm). Taking the

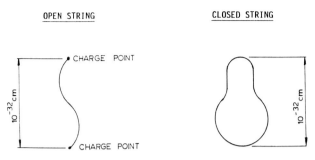

Figure 13.13. Open and closed strings.

Planck energy to be 10^{19} GeV, the tension of the string would have to be m^2 or $(10^{19}$ GeV$)^2$, equal to 10^{39} tons—an enormous value.

It makes the mass scale of particles in the superstring theory infinitely larger than actually observed particles that they have no relation to particles from which the universe is made up: protons, neutrons, and electrons, or even more massive, short-lived particles observed in high-energy accelerators. The particles are infinitely smaller than the masses of particles in the form of strings. We must, however, take into consideration that the superstring theory in ten dimensions has been conceived for the time of the early universe when supersymmetry prevailed. Even so, there are no clear mathematical or physical explanations of how the supersymmetry condition broke up when the universe expanded and cooled down, and how the massive superstring particles broke down to the mass levels of particles now observed.

Even the originators of the theory must admit that the observed masses of basic particles of matter and even those created in accelerators cannot be explained by the superstring theory, as formulated now.

The basic interactions of colliding strings touching, splitting, and joining ends, covered the entire quantum field theory, are shown in Figure 13.14. They are (1) a string splits into two smaller strings, (2) a closed string splits into two smaller strings, (3) two new strings as a result of a collision, (4) an open string changes into a closed and open string, (5) an open string's ends touch and form a closed string.

The lowest vibration mode of a closed string represents a graviton, and the lowest mode of an open string corresponds to a photon. The notion that the superstring theory can unify all four forces of nature comes from the mode of interaction of various strings. The mode of an open vibrating string, which corresponds to a photon mediating the electromagnetic force, can join ends and create a closed string of a vibrating mode corresponding to a graviton, which

Review of Modern Cosmological Theories

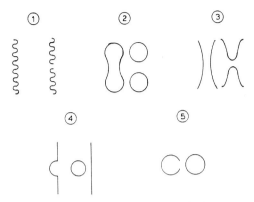

Figure 13.14. The five interactions of strings.

mediates the gravitational force. This phenomenon would then indicate that the particles that mediate the four forces of nature called bosons, originate from one source and can therefore be unified in one theory.

How do strings interact? When two pointlike electrons move and interact, they create a line and exchange a virtual photon in four-dimensional space-time (Figure 13.15). In the superstring theory, the electrons are one-dimensional loops that trace out a cylindrical path in a *ten-dimensional space-time*. The one-dimensional string, when moving, traces out a two-dimensional surface in space-time called a world sheet.

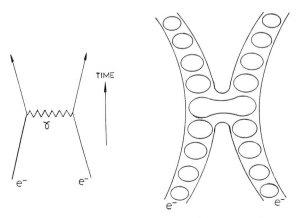

Figure 13.15. Interaction of electrons as particles and as strings.

Figure 13.16. Movement of a point particle, based on general relativity.

13.12.2. Movement of a Pointlike Particle

In general relativity, a point particle moves from an initial position to a final position and traces in space-time a trajectory of minimum length or distance called a geodesic (Figure 13.16). In quantum mechanics, the motion of a particle can take all possible paths; however, the highest probability is given to the path of least action which is closest to the geodesic line of general relativity (Figure 13.17).

13.12.3. Movement of Strings

The string moves also in a way that minimizes its action. *An open string* traces out a world sheet (Figure 13.18). As an open string carries charges at the end points, they define the boundaries of the sheet. *A closed string*, having no end points, sweeps a world sheet that becomes a surface of a minimum area (Figure 13.19).

13.12.4. The Superstring Theory of Creation

In the classical Einstein–Friedmann theory of the birth of the universe from a singularity and the big bang, space as such did not exist. All matter and radiation of the present universe was compressed into zero space or a singularity of infinite density and infinite gravity. Space could not be warped by gravity as it did not exist. Time started with the big bang and therefore to ask what happened before time $t = 0$ and how matter and radiation got into the singularity makes no sense.

Superstring scientists claim that the most spectacular result of their theory is that it can make statements about what happened before the big bang, before the beginning of time itself.

Figure 13.17. Movement of a particle in quantum mechanics.

Review of Modern Cosmological Theories

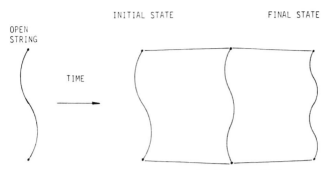

Figure 13.18. Movement of an open string. Result: world sheet.

According to this theory, space originally existed in ten dimensions and in a "false vacuum." As the false vacuum is not at the lowest level of energy, it is unstable. Because of this instability, the ten-dimensional space-time violently ruptured into our own stable four-dimensional space-time and also a six-dimensional space-time, so unstable that it shrunk instantly into six separate small balls of 10^{-33} cm and for all practical purposes disappeared. The splitting was so violent that it created the big bang explosion. The theory goes on to say that when the universe, 10^{-43} sec after the explosion, was only 10^{-33} cm in diameter, matter and energy consisted of unbroken superstrings. This is difficult to understand as a single string, according to the theory, is 10^{-32} cm across and could hardly fit into a smaller space of 10^{-33} cm.

13.12.5. Cosmological Theory of Superstring-Inflation

Another variation of the cosmological superstring theory includes the theory of inflation, which tries to eliminate the problem of the horizon and flatness of the universe in the standard singularity–big bang theory.

In this version of the superstring theory of creation, the violence of the

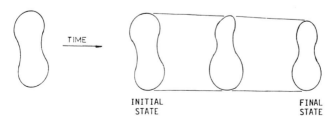

Figure 13.19. Movement of a closed string. Result: A world surface of minimum area.

splitting of the ten-dimensional space generated a temporary rapid expansion of the universe in the false vacuum which is called inflation. The actual big bang explosion took place later on when inflation subsided and the expansion of the universe became normal.

13.12.6. The Free Lunch Cosmological Theory of Superstrings

The superstring theory seems to adapt itself also to the theory of everything from nothing, which proposes that the entire four-dimensional space-time universe was created by a quantum transition from nothing—a pure space-time without matter or energy; a random quantum leap from a vacuum of nothing to a universe triggered by an impulse of a vacuum fluctuation, as described in a previous section.

The violent collapse of the ten-dimensional space-time into a six- and four-dimensional space-time has supposedly given the required impulse for the random leap to create matter out of pure space-time.

While attractive in details and promising in finding solutions, the superstring theory lacks logic and has created more problems than it has solved. This is my opinion after careful study of the claims and resolutions presented by the superstring theory on modern astrophysics. The mathematics of the superstring theory are intriguing and seem to be correct, but what they lack is the relationship between the mathematical formulas and physical reality.

I will list here only a few major deficiencies of this imaginative theory which seems at times to fall into the field of metaphysics and science fiction.

The theory promised to go beyond the big bang and explain the origin of matter. In reality, I found no specific references anywhere as to how matter in the form of strings and radiation became the integral part of a primordial object from which the universe evolved.

The theory has been mathematically formulated in ten dimensions, from which six dimensions curled up in Planck's cells of 10^{-33} cm. There is no explanation as to why, after an explosion similar to the big bang, six dimensions remained curled up and four dimensions unfolded and expanded with the universe. There is also a substantial conflict with the experimentally proven general theory of relativity of curved space-time imposed by gravity of matter, while the superstring theory is based on a flat ten-dimensional space-time where all dimensions are flat.

The highly promoted supersymmetry and unification of the four forces relates to the distance of 10^{-32} cm and high energy of 10^{19} GeV which existed at 10^{-43} sec after the explosion. I have seen no acceptable physical or mathematical method from which low-energy results can be derived and still comply with the presently observable masses of particles and behavior of the four forces of nature.

The most difficult problem, however, remains with the size of the particles of matter in the form of strings. The lightest have zero mass such as photons and gravitons and the next nonzero particle has the enormous mass of 10^{19} GeV, while the measured mass of a proton is 1 GeV or 10 million, trillion times smaller.

There is no acceptable explanation as to how these enormous particles disintegrated into the observable levels after symmetry was broken. One speculation is that matter, made up of those enormous particles sometimes called axioms, exists now as the so-called shadow matter—the unseen mass that makes up the missing part of the critical mass of the universe.

13.13. THE TWISTOR THEORY

Einstein retained the Newtonian idea of space which is continuous and consists of dimensional points. The quantum theory as well accepted space consisting of an infinite number of dimensional points. Many have been puzzled, however, by this principle, which claims an infinite number of points between two points, 1 mm apart, and the same infinite number in the entire universe.

David Bohn suggested that space be described by topology rather than by geometry. This makes sense, especially when space becomes curved by gravity, straight lines become curves, and triangles change into squares or even circles.

Penrose and his associates, on the other hand, developed over the last 30 years an entirely new principle. In the same way that matter has its origin in elementary particles, Penrose claims that space-time has its origin in twistor space. The twistor primordial space had complex dimensions and points were replaced by twistors. The complex twistor space was used to generate our four-dimensional space-time.

Complex numbers are part of a mathematical theory based on the square root of -1 or $\sqrt{-1}$. Minkowski introduced the number $i = \sqrt{-1}$ as the fourth dimension or time in the Einstein four-dimensional space-time. Time t in the Einstein theory of relativity is written as $t_{rel} = ict = \sqrt{-1}ct$. Though time is entered as an imaginary, complex number, the results of a measurement become real numbers.

Space-Time Twistors

If we set up a diagram of space-time where the vertical axis Y represents time or $Y = ict$ and the horizontal axis X the three-dimensions of space, as shown in Figure 13.20, then the distance OP can be determined from the Pythagorean equation

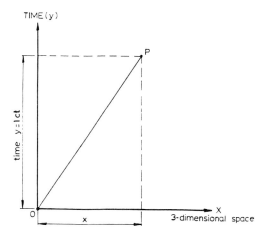

Figure 13.20. Space-time diagram.

$$OP^2 = X^2 + Y^2$$
$$= X^2 + (ict)^2$$
$$= X^2 + (\sqrt{-1}ct)^2$$
$$= X^2 - c^2t^2$$
$$OP = (X^2 - c^2t^2)^{1/2} \qquad (13.18)$$

When an object moves at the speed of light c, $X = ct$ and the distance OP becomes zero:

$$OP = (c^2t^2 - c^2t^2)^{1/2} = 0$$

What we obtain here is a finite line for light, the dimension of which is zero. In relativity, this line of light is called a null line. If, however, a particle moves at a velocity smaller than c, then $X = vt$ and the distance $X_v = (v^2t^2 - c^2t^2)^{1/2} = [t^2(v^2 - c^2)]^{1/2}$. Again, if $v = c$ the space-time length disappears.

If one could travel at the speed of light and measure the elapsed time from O to P, it would be 0. For light rays traveling at the speed of light from a star 1 billion light-years away, the elapsed time to earth is 0, although for us as stationary observers on earth, it takes 1 billion years for the light to reach us.

We have seen that in the superstring theory, which replaces pointless elementary particles with one-dimensional strings of 10^{-32} cm, space-time has been created by or linked to strings.

Another parameter that plays a vital role in the twistor theory is the light cone (Figure 13.21). Light spreads from O with increasing time. Point A is connected;

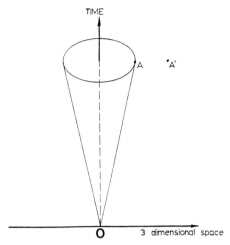

Figure 13.21. Light cone diagram.

however, events at point A' cannot be seen at point O as there is no connection. A' is beyond the light cone. The theory of twistors claims that primordial space prior to any mass could have existed only with geometry based on null lines. Null lines, as explained, are finite lines or tracks left by massless particles such as photons or electromagnetic energy moving with the speed of light. Those lines though finite have zero length. Penrose suggested that the universe started in a space in which electromagnetic energy was moving along null lines, unchanged for eons, until the null lines started to interact and *somehow* mass was created in the universe. He gives, however, no definite solution as to the procedure of mass production. The Penrose primordial space consisted of twistors, massless particles with angular and linear momentum, created out of the straight null lines. This space would then, later, give birth to our four-dimensional space-time.

A twistor, which we can call T, is mathematically defined by complex numbers which are its coordinates and become a point in twistor space. Being a complex number it has the conjugate or mirror-image number T*. It is well known that multiplying a complex number (T) with its conjugate (T*) gives a real number. For example:

$$(1 + 2i) \times (1 - 2i) = 1 - 4i^2$$

as $i^2 = -1$, the result is $= 5$. The product of $\frac{1}{2}T \times T^* = 1$, which is called helicity, is the measure of the twist of twistors. The helicity can be positive, negative, or zero. Penrose divides the twistor primordial space into three regions: The region

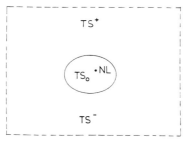

Figure 13.22. Diagram of twistor space.

TS_0 contains twistors with zero helicity $T \times T^* = 0$ and divides the twistor space into two other regions: TS^+ with twistors of positive helicity and TS^- with twistors of negative helicity (Figure 13.22). All points NL in the region TS_0 are equivalent to light rays or null lines in space-time.

14

Vacuum of Space

The classical vacuum of space is considered to be a void, lacking all particles of matter in the form of solids, liquids, or gases and free of all thermal and other radiation—a cold void at absolute zero, $0\,\text{K}$ ($-273°\text{C}$).

Based on the generally accepted big bang theory of creation, space as such outside the singularity did not exist. Space was born after the explosion when matter and radiation expanded, forming together with space the newly created universe. Even today, many leading cosmologists share these views.

When in 1985 I first sent a letter to a select group of astrophysicists and cosmologists outlining the basic elements of my new theory of creation, eliminating the singularity, a famous astrophysicist commented as follows: ". . . there is nothing wrong with your theory but don't describe the expansion of your large fireball as taking place in space. Space as such did not exist before and was being created simultaneously with the expansion. . . ."

Today, there is no doubt that vacuum, even at absolute zero ($0\,\text{K}$) and void of all particles of matter, is far from empty or featureless. Today's theoretical physics, backed up by laboratory experiments extrapolating proven physical laws and applying quantum mechanics, has assigned to the vacuum of space many interesting characteristics.

Some of the remarkable properties that we are going to analyze depend on Einstein's theory of gravitation or Planck's quantum theory; others can best be explained and understood in terms of classical ideas.

14.1. VACUUM FLUCTUATIONS OF SPACE TOPOGRAPHY

The pure vacuum of space prior to Einstein's theory of gravitation was an empty, passive area free of particles, energy, or pressure. Space was an area of the universe independent of the interactions between energy and matter, standing alone and isolated. Einstein introduced the dependence of space and matter. Space tells matter how to move and matter tells space how to curve. For example, gravitation of the sun curves the space around the earth, causing the earth to circle the sun on an orbit resulting from the curved space. Light passing near a large star seems to bend. Actually, it passes through the curved space near the star. It is the matter contained in the star that curved the space around it.

Quantum mechanics introduced another dimension. The vacuum of space, which appears simple and flat to macroscopic observations, is a very complicated and active system in the microscopic world of quantum mechanics. We are speaking here of areas of space of unimaginably minute dimensions in the range of 10^{-32} cm, called the Planck length, 1000 trillion times smaller than an electron (10^{-17} cm). Here, the geometry of the vacuum undergoes dynamic changes. Space in this minute world expands, vibrates, attains maximum dimensions, followed by contraction and collapse. We speak of quantum fluctuations in the geometry and topology of space. This property of being curved, hilly, and distorted is continually passed from one portion of space to another, similar to waves. Empty space is not empty; it is the place of most violent physics. As a result of these fluctuations, matter can appear spontaneously, as described in the following paragraphs.

The vacuum, which looked like an innocent structure for a long time, is rich in physics. It is filled first of all with an enormous number of virtual (to be) particles such as photons and other bosons, pairs of electrons–positrons, muons–antimuons, and others. The virtual particles can be defined as would-be particles that appear spontaneously as vacuum fluctuations, as shown schematically in Figure 14.1. Virtual particles have no mass and appear in the smallest areas of space with a dimension of 10^{-33}–10^{-12} cm. For example, an electron and positron appear as a pair; however, they are annihilated or rather vanish into the vacuum almost as soon as they appear.

They influence the physics at short distances but have no influence on macroscopic physics. This phenomenon can be explained with the Heisenberg uncertainty principle of quantum mechanics, which is expressed in the energy–time equation (ΔE is the uncertainty in energy and Δt in time):

$$\Delta E \times \Delta t = \hbar$$

where $\hbar$ is Planck's constant with a magnitude of 6.6×10^{-22} MeV $\cong 1 \times 10^{-27}$ erg $\cong 1.05 \times 10^{-34}$ J-sec. What happens is that a virtual particle "borrows" a quantity ΔE of energy from the vacuum but must "repay" it before the shortage can be detected. The higher the energy of the virtual particle, the shorter its existence is.

The uncertainty in energy implies that energy conservation can be violated over a very short time, perhaps as little as 10^{-23} sec before it must be repaid. The

Vacuum of Space

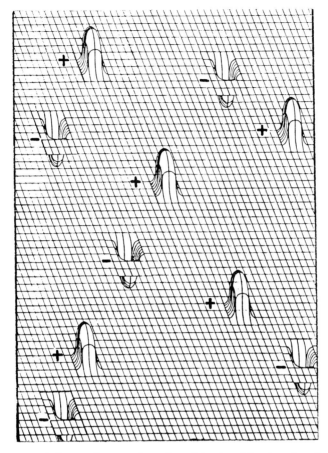

Figure 14.1. Schematic diagram of quantum fluctuations of virtual particle pairs such as electron–positron in vacuum of space lasting 10^{-23}–10^{-15} sec.

total amount of energy remains constant at the 0 level before and after the short appearance of virtual particles, and the energy conservation law is not violated. The existence of virtual particles can only be detected indirectly. If, however, a source of concentrated outside energy is fed into the tiny space of 10^{-11} cm or smaller, equal to at least twice the rest mass energy of an electron of 0.00102 GeV, a pair of virtual electron–positron particles will become a genuine and detectable electron–positron pair. This phenomenon is observed routinely in high-energy particle accelerator experiments where genuine particles leave traces in particle detectors after their creation and before annihilation or decay.

The existence of virtual electron–positron pairs in the vacuum can be

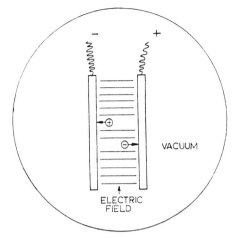

Figure 14.2. Electrically loaded capacitor in a vacuum creates electron–positron particle pairs from virtual particles.

experienced by introducing into the vacuum a sufficiently large electric field in the form of a capacitor or two plates of opposite charge, with an electric field between them as shown in Figure 14.2. This would instantly separate the virtual pair and the virtual particles will become genuine electrons and positrons. In nature, this takes place near black holes where virtual particles appearing from vacuum fluctuations get separated by strong electrical forces and become real particles. Most of the particles fall into the black hole; some, however, will escape outside the horizon creating the impression that the black hole is producing particles of matter. Radiation can also be created as a result of particle and antiparticle annihilation into γ rays of radiation.

Creation of matter and antimatter in the form of an electron–positron pair can also take place in a pure vacuum near a heavy atomic nucleus (Figure 14.4). This works as follows. When a single proton or hydrogen nucleus binds an electron to create a hydrogen atom (see Figure 14.3), energy of 13.6 eV is released

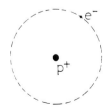

Figure 14.3. An atom of hydrogen.

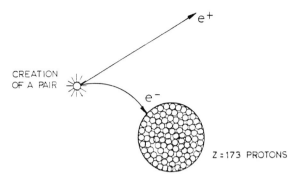

Figure 14.4. Creation of an electron–positron pair from vacuum in the presence of a nucleus of 173 protons.

in the form of light or some other electromagnetic energy. The same applies when an electron is released and the hydrogen atom is ionized. Energy of a minimum 13.6 eV has to be supplied to remove the electron from its orbit. Consequently, the weight (equivalent to energy $E = mc^2$) of a hydrogen atom is smaller by 13.6 eV than the individual proton and electron weights before binding. If a very heavy nucleus could be assembled, a nucleus with $Z = 145$ or $Z = 137$ protons would require a binding energy equal to the rest mass of an electron or 0.5 MeV. The electron would radiate away its entire energy, therefore no mass would be added to the atom.

Such heavy nuclei do not exist in nature, as far as we know, but theoretically they could be artificially created. With $Z = 173$ protons in the nucleus, the binding energy would be equal to two electrons or a pair of an electron and positron. At this level, if a virtual pair of an electron and positron were to appear from a vacuum fluctuation in the vicinity of the nucleus, they would become real particles without the addition of outside energy (Figure 14.4).

The electron would be bound to an orbit and the positron being repelled would escape and could be detected. The neutral vacuum, at this point, would become unstable and become charged.

A heavy nucleus of this type with $Z = 184$ could be created by collisions of two nuclei of uranium $Z = 92$. The heaviest known nucleus of californium of $Z = 98$ could create a nucleus of $Z = 196$.

14.2. INTERACTIONS BETWEEN ELEMENTARY PARTICLES AND VIRTUAL PARTICLES IN THE VACUUM OF SPACE

Under specific circumstances, virtual particles in the vacuum of space interfere with genuine particles such as electrons, affecting their characteristics.

Although the theory of virtual particles was formulated in the early 1950s in a paper by Schwinger, Lamb, and Rutherford, already in 1947 the interaction of virtual particles with electrons rotating around the proton in hydrogen atoms was noticed. The constant short appearance and disappearance of virtual pairs create intermittent short-lived electric fields, which cause the electron to jiggle in its orbit. This phenomenon gave strong support for the idea that all space in the universe is full of virtual particles and antiparticles.

14.2.1. Scattering of Electrons

An interesting experiment is the scattering of two high-energy electrons in the vacuum, causing vacuum polarization. At normal distances larger than 10^{-11} cm, electrons will scatter and the repelling force will comply with Coulomb's law. The force is proportional to the square of the electrical charge of the electron. However, when we measure the force between two high-energy electrons that come closer than the critical distance of 10^{-11} cm, the forces between the electrons are larger than those calculated on the basis of Coulomb's law. This phenomenon can be explained as follows.

Space around an electron is filled with virtual positrons that influence the distribution of the electrical charge of the electron. What happens is that the negatively charged electron repels all virtual electrons and attracts virtual positrons, as shown schematically in Figure 14.5. The cloud of virtual positrons surrounds the electron, causing the vacuum around the electron to become polarized. Such an electron has a partially shielded charge and is called a physical or "dressed" electron. An electron without the virtual positron cloud is termed naked. At normal levels of electron energy, the effect of the virtual positrons is difficult to detect. In the case of the experimental scattering of high-energy

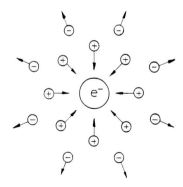

Figure 14.5. A high-energy electron attracts virtual positrons from the vacuum of space.

Vacuum of Space

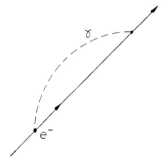

Figure 14.6. Balancing of electron's energy by emittance of virtual photons (γ).

electrons, the electrons become undressed. As the charge of a naked electron is greater than that of a physical electron, the force with which the electrons repel each other under normal conditions no longer follows Coulomb's law.

14.2.2. Emission of Virtual Photons

Another example of the uncertainty principle effect and the interaction of virtual particles in the vacuum of space is the continuous emittance and absorption of virtual photons by electrons, balancing their energy (Figure 14.6). The electron is surrounded by a cloud of virtual photons, schematically shown in Figure 14.7. As photons have 0 rest mass, any slight fluctuation in the energy of the moving electron triggers the appearance of virtual photons, which are always ready to act and disappear. It is a sort of continuous interaction between the electron and vacuum of space.

14.2.3. Scattering of Two Electrons

Another example of interaction of virtual photons is the scattering of two close-passing electrons, which are normally depicted by a Feyman diagram showing the exchange of a virtual photon and the electron's direction (Figure 14.8).

Figure 14.7. Electron surrounded by a cloud of virtual photons.

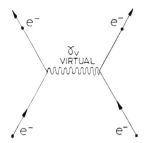

Figure 14.8. Scattering of two electrons.

Considering the fact that the electrons are surrounded by a cloud of photons, the exchange of the virtual photon between two interacting electrons can be explained more clearly as an interaction of the two virtual photon clouds with each other. In other words, some of the virtual photons move across from one cloud to the other (Figure 14.9).

14.2.4. Gluon Clouds

Quarks, which are firmly confined in a proton or neutron, are also surrounded by a virtual gluon cloud. In high-energy proton–proton cyclotron collisions when protons come very close to each other, an exchange of virtual gluons from the gluon clouds takes place.

14.2.5. Virtual Particles Become Real near Black Holes

Hawking came up with a theory that near tiny primordial black holes, or rather in the vicinity of the horizon, virtual particles become real ones. In this theoretical process, when a virtual pair appears and separates for a moment, one

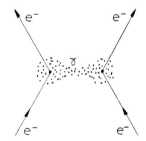

Figure 14.9. Scattering of electrons by exchange of virtual photons.

particle is absorbed by the gravity and eaten up by the black hole, while the other becomes a real particle in the real universe. The theory goes on to say that the energy provided to the virtual particle comes from the gravitational field energy of the black hole. As the black hole gives away energy, it is gradually losing it ($E = mc^2$) in the process of final evaporation. Hawking continues by saying that for every kilogram of particles created, the black hole loses its mass by 1 kilogram. There is a lot of scientific contradiction in this reasoning.

First of all, a black hole is supposed to create around its singularity an infinite curvature of space, and gravity is so strong that there is no possibility for created particles to escape it.

It is also my opinion that in such a situation, the created particles would borrow the energy from the gravitational field but would be "eaten up" as soon as they are created, balancing out the total energy content of the holes.

Hawking also speaks about the decreasing mass of the black hole, while the theory claims that all matter has been squeezed out of existence during the implosion of mass and the black hole singularity is empty.

14.2.6. False Vacuum

When outside, high energy is put into the vacuum resulting in very high temperatures, as was the case during the birth of the universe, the vacuum acquires interesting properties and actually becomes a "false" vacuum. According to this theory, the false vacuum does not get less dense, when it expands, as anything else that expands, such as matter. The energy is flowing in and keeps the density constant.

According to Einstein's theory of relativity, the expansion of the universe in the early stages, and even today, depends on the energy density. The average density of particles, for instance, decreases as the universe expands, which slows down the expansion rate. Consequently, if the expanding universe were to hit a false vacuum where the density remained very high and constant for a period of time, the universe would suddenly expand at a considerably higher rate, like a giant balloon. This phenomenon is the basis for the cosmological theory, called inflation, which accepts the singularity–big bang concept but adds to it a short period of sudden and enormous expansion with a factor of 10^{30}–10^{40} and which was supposed to have taken place at 10^{-35} sec after the big bang. Still following this theory of the false vacuum and inflation period, after a short while the vacuum energy would decay to a value it would have had without being a false vacuum and the inflationary expansion would stop. Guth, who initiated the inflation theory, gave the following explanation for this phenomenon.

As the contents of the early expanding universe of photons, quarks, and electrons were cooling down, the energy of the vacuum of space was cooling down as well. However, at approximately 10^{-35} sec after the big bang when the universe was as small as a proton (10^{-13} cm), suddenly the energy density of the vacuum got "hung up" at the same value that would normally correspond with the rate of expansion and cooling down, and this caused the sudden superexpansion or inflation of the universe. After a short period of 10^{-5} sec, the density of the vacuum decayed to a normal value, inflation stopped and the universe continued to expand and cooled down at a normal rate, as did the density of the vacuum. More detailed information about inflation is contained in the analysis of cosmological theories of creation.

14.2.7. Vacuum Is Filled with Fluctuating Electromagnetic Fields Called Zero-Point Radiation

It has been established experimentally that when all matter is evacuated from the vacuum of space and all thermal electromagnetic radiation dissipates to nil when the temperature is cooled to 0 K, a residue fluctuating electromagnetic field remains in the vacuum. This electromagnetic field, called zero-point radiation, is not the blackbody background radiation found everywhere in the universe as the remnant radiation of the primordial fireball during the creation act.

The discovery of the zero-point radiation in the vacuum of space has implications regarding my own theory of creation, described later. It is, therefore, important to review the findings of laboratory tests on this subject.

14.2.8. Thermal Radiation

Under conditions of equal temperature, thermal radiation is homogeneous and isotropic, equal in every point and direction of space. The temperature of thermal radiation determines not only the energy density but also its spectrum, or the amount of energy at various frequencies of waves.

14.2.9. Energy Density of Radiation

The Stefan–Boltzmann equation of total energy density of thermal radiation is $E = aT^4$ and indicates that at $T = 0$ K, the energy of a vacuum also falls to 0. This condition described the classical vacuum, a space of nothing, no particles, no energy, no temperature.

If E is the total energy radiated at all frequencies, we proceed as follows. The number of waves passing through a given point in 1 sec is called the frequency v. The distance λ shown in Figure 14.10 between the same points on the waves is called the wavelength. Electromagnetic waves propagate with the electric and

Vacuum of Space

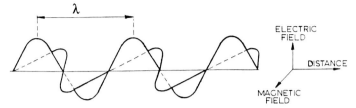

Figure 14.10. Propagation of electromagnetic waves.

magnetic fields at right angles. As the velocity of all electromagnetic waves is c, in time t, vt waves are emitted and they occupy a length of $vt\lambda$:

$$\frac{vt\lambda}{t} = v\lambda$$

or

$$c = v\lambda$$

$$\text{Energy } E = \frac{\hbar c}{\lambda}$$

The shorter the wavelength, the higher is the energy of radiation. However, energy also depends on the temperature of the radiating source (Figure 14.11).

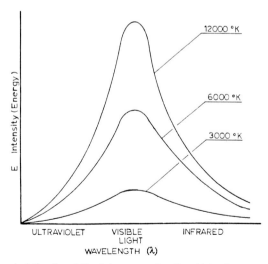

Figure 14.11. Diagram depicting the relation between wavelength and intensity or energy E of blackbody radiation at 3000 K, 6000 K, and 12,000 K.

14.2.10. Blackbody Radiation

$$\text{Energy density} = aT^4$$

where $a = 7.5647 \times 10^{-15}$ erg cm^{-3} K^{-4}

$$\text{Emittance} = \sigma T^4$$

where $\sigma = 5.6696 \times 10^{-5}$ erg cm^{-2} sec^{-1} K^{-4}.

p is the momentum, E is the energy, and λ is the wavelength

$$p = \frac{(E^2 - m^2c^4)^{1/2}}{c}$$

m for photon $= 0$

$$p = \frac{E}{c}$$

$$\lambda = \frac{\hbar}{p} = \frac{\hbar c}{(E^2 - m^2c^2)^{1/2}} = \frac{\hbar c}{E} \qquad \begin{aligned} E &= pc \\ E &= \frac{\hbar c}{\lambda} \end{aligned}$$

$$v = \frac{pc^2}{E} = c\left(1 - \frac{m^2c^4}{E^2}\right)^{1/2} = c$$

The higher the temperature, the higher is the energy density of radiation. Planck found the relation between the radiant energy lv, wave frequency, and temperature:

$$lv = \frac{2\pi h}{c^2} \frac{v^3}{e^{hv/kT} - 1}$$

Total energy with contribution of all wavelengths from zero to infinity:

$$E = \int_0^\infty lv \, dv$$

or

$$E = \frac{2\pi h}{c^2} \int_0^\infty \frac{v^3 \, dv}{e^{hv/kT} - 1}$$

Integration over λ from 0 to ∞ gives an energy density formula already known:

$$E = aT^4$$

If we calculate the blackbody radiation of the original fireball containing the universe at today's temperature of $T = 2.7$ K, we get

$$E_{BBR} = a \times (2.7^4) \quad \text{or} \quad 4 \times 10^{-13} \text{ erg/cm}^3$$
$$1 \text{ eV} = 1.6 \times 10^{12} \text{ ergs}$$
$$E_{BBR} = 0.25 \text{ eV/cm}^3$$

which is approximately equal to the radiation density of the Milky Way galaxy. The calculated peak is at a wavelength of $0.2(hc/2.7 \text{ K})$ or 1 mm.

14.2.11. The Zero-Point Electromagnetic Radiation

In 1958 M. J. Sparnaay proved experimentally that the vacuum of space even at 0 K, without any particles or thermal radiation has an inherent fluctuating electromagnetic radiation field, which he named zero-point radiation. This surprising electromagnetic field has been discovered during a classical Casimir experiment with two electrically conductive plates in a perfect vacuum when an attractive force was created between the two plates, as shown in Figure 14.12. Measurements determined that the attractive electric force F_A is proportional to the area A of the place and inversely proportional to the fourth power of the distance, with constant $a = 1.3 \times 10^{-18}$ erg-cm:

$$F_A = a \frac{A}{d^4} \tag{14.1}$$

In the specific Sparnaay experiment, the plates had a 1 cm² area and were separated by $d = 0.5$ μm. The Casimir attractive force F_A was equivalent to 0.2 mg.

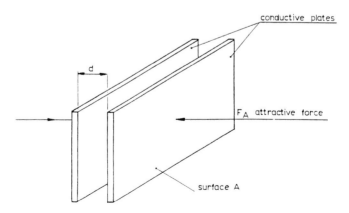

Figure 14.12. An attractive force F_A has been created in a pure vacuum by the zero-point electromagnetic radiation.

In a similar Casimir experiment with thermal radiation, the observed attractive force is directly proportional to temperature and indirectly proportional to the cube of distance:

$$F_{th} = a\frac{T}{d^3} \qquad (14.2)$$

At $T = 0\,\text{K}$, force $F_{th} = 0$.

14.2.12. Characteristics of the Zero-Point Radiation

1. The zero-point radiation, similar to thermal radiation, is homogeneous and isotropic.
2. Being a radiation field in the vacuum it must look the same way to any two observers, moving with different velocities, provided the velocity is constant. Surprisingly, however, the zero-point radiation is not affected by the Doppler effect.

The velocity of all electromagnetic radiation in a vacuum is constant and equal to 2.99793×10^{10} cm/sec, regardless of the different speeds of moving observers. Each observer finds the same velocity of light. However, the wavelength changes if the source of electromagnetic radiation and the observer are in relative motion. This is called the Doppler effect, described in detail in the chapter dealing with the expansion of the universe (Figure 14.13). In Figure 14.13, the source S is emitting waves at ν frequency and λ wavelength. If the observer O is stationary and the source S is moving with velocity v (cm/sec), then in time t (sec) the νt waves emitted by S will cover a distance $\nu t \lambda + vt$ instead of $\nu t \lambda$.

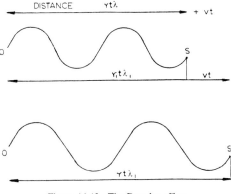

Figure 14.13. The Doppler effect.

Vacuum of Space

As the same number of waves occupy a larger length ($+vt$), the observer sees an increase in wavelength λ. As the velocity must be the same, frequency must decrease:

$$vt\lambda_1 = vt\lambda + vt$$

Dividing by vt and since $c = \nu\lambda$

$$\lambda_1 = \lambda + \frac{v}{\nu} = \lambda\left(1 + \frac{v}{c}\right)$$

As speed c must be equal

$$c = \nu_1\lambda_1$$

or

$$\nu_1 = \frac{c}{\lambda_1} = \frac{\nu}{1 + \frac{v}{c}} = \nu\left(1 + \frac{v}{c}\right)^{-1}$$

If the source S is approaching the observer O, the vt waves must be reduced to a shorter space and the observer will see a decrease in the wavelength.

If the electromagnetic energy is light the spectrum would move to blue, but if the source is moving away from the observer the spectrum would shift to red. It turns out that the intensity of the zero-point radiation at any frequency ν is proportional to the cube of the frequency ν (Figure 14.14):

$$I_{ZPR} = \text{constant } \nu^3 \quad \text{ergs sec}^{-1} \text{ cm}^{-2} \text{ Hz}^{-1} \text{ steradian}^{-1}$$

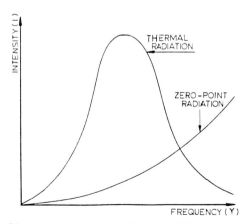

Figure 14.14. Diagram comparing standard thermal radiation with zero-point radiation.

From the Sparnaay experiments, the constant is equal to 3.3×10^{-27} erg-sec or half of Planck's constant. Therefore,

$$I_{ZPR} = \tfrac{1}{2}\hbar v^3 \qquad \text{ergs sec}^{-1} \text{ cm}^{-2} \text{ Hz}^{-1} \text{ steradian}^{-1} \qquad (14.3)$$

The randomly fluctuating electromagnetic field with the zero-point spectrum is not subject to the Doppler effect due to compensating changes in frequency and intensity. Increase in frequency is compensated by increase in intensity for approaching observers and vice versa for a receding observer.

14.3. THE HIGGS VACUUM FIELDS AND THE HIGGS BOSON

This theory, developed by P. W. Higgs, assigns to the vacuum an altogether new property. It claims that the vacuum of space contains a constant energy field called the Higgs field, which can couple with massless particles such as bosons W^+, W^-, and Z^0, responsible for the weak interaction, and give them mass.

As we have already established, the basic parameters of modern physics include not only the particles themselves but the four forces of nature, which, together with the particles, can all be called fields.

The concept of a field is a quantity defined at every point of a certain region of space and time. The electromagnetic, weak, and strong nuclear force fields are vector fields. This means that at each point of space where the field extends its presence, one can draw a vector with its magnitude expressed by the length and its direction, which in a three-dimensional space can be determined by two angles. Three numbers are therefore required to specify the value of the vector: magnitude plus two angles. The coupling boson particles of those fields which are responsible for all interactions of particles of matter—the photon, W^+, W^-, and Z^0, and the gluons having a spin 1—are vector fields. The gravitational field is a tensor field and has ten components. The quantum of the field or the coupling boson is the graviton with a spin of 2, corresponding to five spin states, as shown in Figure 14.15.

An electromagnetic field expresses the way the force of electromagnetism is transferred from one place of space to another. All charged particles create around themselves an electromagnetic field (Figure 14.16). Each particle interacts then with the sum of all of the other electromagnetic fields when particles move. In a static situation, the field is electric only.

In quantum mechanics the particles themselves are represented as fields. An electron, for instance, is considered as a quantum of waves extending into space. If two electrons, each surrounded by its electromagnetic field, approach each other, they exchange a virtual photon or quantum of electromagnetic field. The virtual photon which cannot be detected, once emitted, is absorbed. The greater its energy, the shorter is its existence.

Vacuum of Space

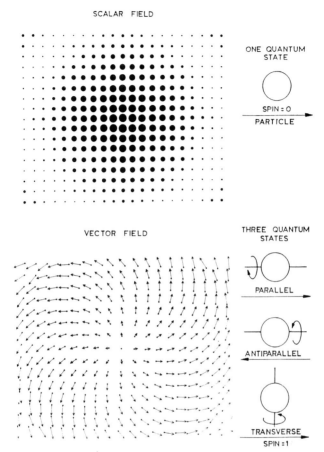

Figure 14.15. Energy fields. A scalar field has a magnitude only at each point. A vector field has a magnitude and direction. A scalar field may represent temperature or density; a vector field the earth's velocity. The quantum of the scalar field has only one component, one spin state; the three-dimensional vector has three spin states. (Adapted from "The Higgs Boson" by Martinus J. B. Veltman, copyright *Scientific American*.)

Each field has its mediating particles, called bosons: the photon for the electromagnetic field, three weak vector bosons W^+, W^-, and Z^0 for the weak nuclear field, eight gluons for the strong nuclear field, and the graviton, a tensor (not yet detected), for the gravitational field.

Higgs introduced his complementary energy field claiming that the standard model of particle physics with the four fields is incomplete, especially when trying to explain the heavy mass of the W^+, W^-, and Z^0 bosons.

The Higgs field is assigned a unique property and is supposed to be present in all vacuum of space. The basic characteristic of the vacuum, as defined in

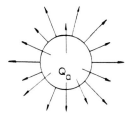

Figure 14.16. The electrical field of charged particle Q_a extends into space in three dimensions.

classical physics, calls for all known fields to have a zero energy level. For example, an electron field reaches its zero value when there are no electrons.

In the case of the hypothetical Higgs field, however, in order to reduce its energy level to zero, new energy input is required and its lowest value in the vacuum is a uniform level, greater than zero. The theory, which is not substantiated by experimental evidence, claims that the entire vacuum of outer space contains the Higgs constant and uniform energy field, larger than zero.

In addition, the Higgs field, in contrast to the described vector field of the three forces of nature, is a scalar field, which means that each point has a magnitude (not direction) and the quantum of the field has a spin of zero. In an analogy with the mediating particles of the three known vector fields and the tensor gravity field of the standard model of elementary particles, the Higgs field has its own mediating particle known as the Higgs boson with spin 0. The theory further claims that the Higgs field can generate mass by coupling with certain particles. The stronger the coupling, the greater is the mass of the particle.

As already mentioned, the Higgs scalar field is represented by a spin-0 particle or scalar boson with one spin state—the Higgs boson. Vector fields mediate by spin-1 particles or bosons with three spin orientations. This applies to the photon, W^+, W^-, Z^0, and gluons.

As explained earlier, the original theory of the weak force, as developed by Glashow, implied that the W^+, W^-, and Z^0 mediating bosons of the nuclear weak force energy field would be massless, similar to the photon, which has no mass. The fact, however, that the weak force has an extremely small range of activity, only up to 10^{-15} cm or approximately 100th of the radius of a proton, implied that the virtual particles W^+, W^-, and Z^0 exchanged in weak interactions must be massive.

Abdus Salam and Steven Weinberg independently suggested that the weak interaction bosons $W^\pm$ and Z^0 are massive particles, which subsequently was proven in many experiments at CERN. The W and Z bosons are very unstable and their lifetime is 10^{-23} sec. They decay into electrons, positrons, and neutrinos.

It is not clear why nature uses the so-called Higgs field in acquiring mass to the weak force bosons $W^\pm$ and Z^0 and not to photons, which are massless.

15

The Velan Multi-Universe Cosmos, the First Cosmological Model without Singularity

All presently published cosmological theories, even those that try to amend the classical singularity theory as the birthplace of the universe by speculative mathematical formulas and principles, assign an absolute nothingness to the time before the big bang explosion, a void. No space, no energy, no dimension, no beginning, no time, no eternity; a mathematical empty set, a nothingness without laws of behavior.

It is difficult to believe that the Creator would choose such a void of nothingness as His environment and then suddenly, only 18 billion years ago, order the mathematical point called the singularity to transform itself into a magnificent universe, full of glory and splendor. A universe, intricate and ingenious, with vast structures of space-time, galaxies, stars, pulsars, quasars, black holes, the mysterious world of elementary particles and their interactions, the complex reality of living cells, and the enormous variety of creatures, all governed to the smallest detail by well-defined laws of nature. A complicated world of texture and color evolved from the interaction of simple elementary particles of matter with energy.

Many think that to ask what happened before the big bang makes no sense, as stated typically by the Soviet physicist I. M. Chalotnikov: "As the beginning of time is simultaneous with the creation of the universe, the question makes no sense" and that the particle creation process does not even belong to science but rather to metaphysics or religion.

God, being capable of doing anything, could have initiated the universe in any way He wanted, even in concentrating all mass and energy in a mathematical point and then letting it explode; all this in violation of established physical laws, but then He could also have chosen not to allow the universe to evolve after birth. The thought that God established only one universe would put considerable limitations on His creativity.

An extensive analysis of these thoughts over a long period of time led me to the conclusion that the moment of creation must have been governed by the rules and physical laws of the cosmos that apply in our universe today, perhaps even without outside intervention. It is this philosophy that guides my new theory of creation. A universe born in a dark and cold cosmos of immense vastness, illuminated by other universes in various stages of development, some just being born and others in full development, contracting to their death. Such an overall theory of creativity seems to me a much more acceptable and complete assessment of the Creator's potential.

Though the size of our universe and its vast activities are mind-boggling, the restrictions put on the creativity potential of God by limiting the cosmos to a single universe make little philosophical sense. The idea of a multiple-universe cosmos is not new. In 1714, G. W. Leibniz, the famous German philosopher, wrote in his *Monadologia*, ". . . There is an infinite number of possible universes. . . ." There has, however, been no meaningful, scientific follow-up of the idea so far.

A few words on how a new theory in science is sometimes developed: Though the basic idea of a new scientific theory may come in the form of a sudden vision in a single stroke, the idea itself must be developed in an incremental, critical thinking process and with a special type of determination and concern; all based on a thorough knowledge of all background scientific data.

What appears to be a sudden solution to an old problem is actually the result of long, creative concern, detailed thoughts, incremental changes, and deeply critical, step-by-step evaluation—all finally crystallizing into a new theory.

In my particular case, the basic idea that, I think, is a big leap forward toward the total solution of the mystery of creation came under most unusual circumstances. One evening in January 1985, I watched in my Florida home the first film presentation of a horrifying story of a hydrogen bomb holocaust called "The Day After." A scientific and political analysis and discussion followed the presentation. When I finally retired after midnight, I suddenly became aware of a unique experience. My brain seemed to clear in a flash from all interference, ready to receive a message. In a split second the basic idea of my cosmological theory crystallized. What followed is nearly seven years of consistent, critical evaluation and analysis of the basic theory. In other words, the full theory had been worked and reworked mentally long before any words were put on paper.

My theory of creation is so far the most complete story ever told of how particles of matter were created and where the enormous energy contained in the universe came from. All processes described are in full compliance with the physical laws as we know them and as they have been proven in laboratory experiments or by astronomical observations.

My theory is the only cosmological theory that eliminates the mysterious and unscientific singularity as the only source of creation. In addition, no one else has succeeded in describing events prior to 10^{-45} sec after the big bang.

Steven Weinberg, in his famous book *The First Three Minutes*, describes the new universe only when the temperature had cooled down to 100,000 million K (10^{11} K) from 10^{25} to 10^{32} K at the time of creation. Our ignorance of microscopic physics, claims Weinberg, at temperatures higher than 10^{11} K obscures our view of the very beginning.

SIX POSTULATES FORM THE BASIS OF THE VELAN THEORY OF CREATION

Six postulates form the scientific basis for my new theory of creation. A short review is planned here of all six postulates focusing on the most important features and establishing a solid foundation for the new theory.

Postulate 1: The Cosmos

The Velan cosmological theory introduces a completely new environment in which our universe was born as part of a cosmos, the home of an indefinite number of universes.

The cosmos existed for eons of time, vast and mighty, infinite in space-time, and an everlasting witness to the glory of the Creator. The cosmos contains many universes with their own initial configurations and at different stages of development. They illuminate the profound darkness of the cosmic space, some appearing as bolts of light, like stars, others shine like mighty galaxies, and some are invisible. If one could stand 30 billion light-years away, our universe would appear as a round galaxy.

The universes, created at different cosmic times, are filled with matter and radiating energy, perhaps similar in structure to our universe. Some just being born, others at the end of their evolution, still others young and active, full of stars and galaxies, all guided by the same rules and physical laws from their birth, their expansion through their own horizon, determined by its total mass, its evolution, its gravitational implosion, and finally explosion into a new cycle of expansion and development.

The radius of the horizon, or a spherical, maximum space in which a given universe can expand and no matter or radiation can escape, can be calculated from the Schwarzschild equation, which also applies in the cosmos:

$$r_M = \frac{2GM}{c^2}$$

where G = the gravitational constant, M = the mass of a given universe, and c = the speed of light. For our own universe, with the mass of 5.68×10^{56} g, the radius is 89 billion light-years.

The various universes are all self-contained units and do not influence other universes. As conditions to develop complicated living organisms on a planet such as the Earth must be extremely fine-tuned, intelligent life, in particular, must be rare throughout the cosmos. When we speak of fine-tuning, it includes the original state of a particular universe at creation and its exact rate of expansion. Any change in parameters can cause an early collapse, long before life can begin anywhere. Then, an early generation of stars must form to convert the primordial light elements of hydrogen and helium into such heavier elements as carbon and oxygen, out of which we are made.

The first generation of stars must explode in supernova fashion to provide the "dust" of heavy elements for the second generation of stars such as our own sun and planetary system. It takes a minimum of a few billion years of evolution to achieve this state of development.

The cosmos contains also a four-dimensional space-time vacuum which, similar to our own universe, is a rich environment, dense with virtual particles, probing to become real and an active space topography in the small space cells of 10^{-33} cm or superspace. The physical laws which govern our universe are also dominant in the cosmos—a logical conclusion to the work of the Creator.

Postulate 2: The Four-Dimensional Cosmic Space-Time

Cosmologists who believe in the classical singularity–big bang theory of creation or in amended versions, such as inflation or superstrings, claim that space was created simultaneously with time, matter, and the universe when the singularity exploded into a big bang.

To ask what was before is a meaningless question as there was no before. It is an inpenetrable mystery, beyond human reach. There was no time, no space, only nothingness. In my theory of creation, the cosmos, which contains an indefinite number of universes, extends over an enormously vast, maybe infinite, cosmic interuniverse space. The vacuum of this cosmic space has similar characteristics as our own. It is dense with virtual pairs of particles, dynamically active in the tiny cells of superspace, and is curved in the vicinity of large masses of matter.

The Creator has assigned to space a major role in creation and we must logically assume that space has existed in the cosmos long before even the first universe was born.

Postulate 3: The Vacuum of Space is Dense with Virtual Particles, "Seeds" of Matter

We have discussed in extensive detail the vacuum of space in Chapter 14.

The classical vacuum of space which used to be considered in the past an empty void of nothing is actually an enormously rich environment, dense, with invisible, virtual, or photons to be and particle pairs of electrons–positrons, quarks–antiquarks, and other matter. They make their appearance in extremely short-lived (10^{-10} to 10^{-23} seconds) vacuum fluctuations and disappear without being directly detected. The heavier particles already described seem to "borrow" energy from the vacuum and make their appearance, but return the borrowed energy and disappear before detection can take place, seemingly avoiding violation of the conservation law of energy and electric charge. The short time violation of the law of mass-energy conservation is allowed by the Heisenberg uncertainty principle and the time can be calculated:

$$\Delta t \times \Delta E = \hbar$$
$$\Delta E = 2 \times 0.511 \text{ MeV}$$
$$\Delta t = \frac{\hbar}{\Delta E} = 10^{-10} \text{ seconds}$$

In case of a pair of electron positron with total rest mass or energy ΔE, the short-lived time is 10^{-10} seconds.

In the presence of an outside source of a highly concentrated electromagnetic energy field in the form of γ rays, equal for instance to at least the rest mass of an electron-positron pair, the virtual particle pair becomes a real or true electron and positron pair. What takes place is an instantaneous, direct transformation of electromagnetic energy into matter, following the equivalence formula of Einstein $E = mc^2$ or $m = E/c^2$. The mass to the particles was given by the outside energy field.

In quantum mechanics, the potential density of virtual particles in the vacuum of space is enormous. It is estimated at 10^{94} g/cm^3. This compares with a density level of 10^{14} g/cm^3 called nuclear density when elementary nuclear particles are squeezed to the level of an atomic nucleon.

The density of virtual particles in the vacuum of space represents a potential capability of producing electron–positron pairs as follows:

one electron–positron pair weighs 1.82×10^{-27} g

1 cm^3 of space could, therefore, potentially produce

$$\frac{10^{94}}{1.82 \times 10^{-27}} = 10^{94} \times 0.54 \times 10^{27} = 5.4 \times 10^{120} \text{ of electron–positron pairs}$$

The entire universe, however, contains only 3.05×10^{80} electrons or 10^{41} less than potentially 1 cm³ that space could produce. The theoretical or potential transformation capacity of virtual electron–positron pairs and other particles into real particles would require the presence of an enormously powerful concentrated energy source in the form of γ rays. Let's calculate what would be required.

One electron–positron pair has a rest mass of 1.022 MeV or 1.022×10^{-6} GeV.

Therefore,

$$\text{the total energy required would be } 5.4 \times 10^{120} \times 1.022 \times 10^{-6} \text{ GeV}$$
$$\text{or } 5.51 \times 10^{114} \text{ GeV}.$$

Such an unimaginably high and concentrated energy level is even theoretically impossible to achieve. However, an energy level of 10^{12} GeV available in a space of 1 cm³ could transform at least $10^{12}/(1.002 \times 10^{-6}) = 1.002 \times 10^{18}$ virtual particle pairs into true electron-positron pairs.

As we can see, particles of matter in the universe represent, percentagewise, a negligibly small part of the enormously large vacuum potential as a source for particle creation. The vacuum of space is always ready to be transformed into particles of matter in the presence of powerful energy fields. Therefore, it seems logical that this phenomena was certainly an essential ingredient in the creation process of our universe in the cosmic vacuum of space.

Postulate 4: Large Fluctuations of Topography Take Place in Tiny Cells of Cosmic Space

It has been known for over 25 years that, in addition to the Einstein gravitational effect or curving of space in the presence of matter, as discussed in Chapter 10, extremely small space cells in dimensions of 10^{-35} cm or a billion, billion times smaller than an electron (10^{-17} cm) undergo violent dynamic changes in their geometry. These tiny sections also called superspace, which may even contain more than four space-time dimensions, undulate, vibrate, expand, attain maximum dimensions, collapse, and explode. The fluctuations in the geometry and topology of superspace to which ordinary laws of geometry do not apply coexist with the other type of vacuum fluctuations which cause virtual particles to appear and disappear.

The distortions and fluctuations are passed on from one area of space to another, similar to waves. This quantum-mechanical characteristic of space plays a major role in my theory of creation. Many scientists, including Sakharov, tried to somehow combine these fluctuations in superspace, also called pregeometry, with a particle-creation process and interactions. This would have been a simple solution but, so far, no detailed ideas or procedures have been presented by anyone to resolve the speculations.

The Velan Multi-Universe Cosmos

Figure 15.1. Quark–gluon plasma. This picture, photographed by R. Stock at CERN, Geneva, is a collision of ultrarelativistic sulfur ions against stationary gold nuclei. It is the first in history, re-creating conditions of the universe in the primordial fireball, as depicted in the Velan cosmological model when matter was so hot and dense that quarks and gluons that now make up nuclear matter were unconfined. The resultant particle shower is recorded in a streamer chamber. Annihilation energy was so great that a plasma of quarks and gluons was created, a small fireball, which lasted 6.5×10^{-23} sec. Free quarks combined into hadrons after the fireball expanded to 6.5–8×10^{-13} cm. (CERN)

Postulate 5: Free Quarks at Creation Time

When I first discovered my cosmological theory in January 1985, it was generally predicted that although all quarks in the universe are confined to protons, neutrons, other baryons, and mesons, at very high field temperatures above 10^{13} K, combined with high densities which prevailed during the creation period, quarks existed as free particles.

As mentioned earlier in this book, it has been proven in accelerator experiments that the force between quarks seems to disappear altogether when the quarks are squeezed close to each other at distances less than 10^{-13} cm. During the birth of the universe, the densities, temperatures, and energy levels were so high that the strong nuclear force was united with the two other forces of nature and, therefore, all particles were free, subjected to gravity only. After the universe cooled down sufficiently, as we will see later, all free quarks were confined in protons and neutrons. Free quarks play a substantial role in my theory of creation. It is, therefore, extremely satisfying that a group of scientists at CERN in the years 1987 to 1988 re-created, partially, the conditions of the universe just seconds after creation when matter was extremely dense and hot. Ultrarelativistic ions of sulphur were smashed in high-energy accelerators against stationary ions of gold, 32 projectiles of sulfur against 80 nucleons of gold, with an energy of 20 GeV per nucleon.

As shown in Figure 15.1, a quark–gluon plasma "fireball" was created from high-energy (γ) photons resulting from annihilation of the particles of gold and sulphur. The "fireball" or miniature universe was almost spherical, slightly flattened along the collision axis. The plasma was extremely dense and lasted only 6.5×10^{-23} seconds. Quarks were obviously prevented from binding when density was so high that the range of the interquark gluon force was less than the radius of a proton or 10^{-13} cm. The tiny "fireball" expanded rapidly to about 6.5–8×10^{-13} cm before it "froze up" into individual hadrons such as protons made up of confined quarks.

Similar tests continue and will inevitably fully confirm the theory of free quarks at high-energy, temperature, and density levels that prevailed in the early universe.

16

The Cosmic Primordial Electromagnetic Radiation Field
Postulate 6

The concept of an omnipresent primordial, electromagnetic, radiation field flowing through the interuniverse cosmic space at velocities greater than the speed of light came to me in a visional flash, combined with the overall theory of particle creation and subsequent events resulting in the birth of our universe. This powerful, basic energy field of the cosmos in the form of super-high-frequency radiation, at energy levels of maybe 10^8–10^{11} GeV, is the fundamental tool of divine power. It propagates at velocities largely exceeding the speed of light of $c = 299,729$ km/sec. This statement is really not a contradiction of the Einstein theory of special relativity but rather a limitation put on the theory and its validity to those parts of the cosmos where the pure cosmic vacuum is replaced by a mixture of space matter and radiation as it exists in our own universe.

The primordial radiation field is the "missing" link in making any theory of creation viable. It is the basic energy field of the cosmos which, together with the vacuum of cosmic space, carry jointly all the seeds of matter of the four forces of nature responsible for particle interactions, as well as the "chromosomes" of all physical and biological laws. All are unified, waiting to unfold their individual identities at different levels of energy, temperature, and optimum chemical and physical conditions. It took, for instance, 10 billion years for life to develop in our area of the universe.

The cosmic, primordial electromagnetic radiation field can interact occasionally with the inherently rich environment of the vacuum of space and turn, in combination with violent fluctuation of superspace, virtual particles into matter. A pure transfer of energy into matter. Electromagnetic radiation at various wave-

THE ELECTROMAGNETIC SPECTRUM

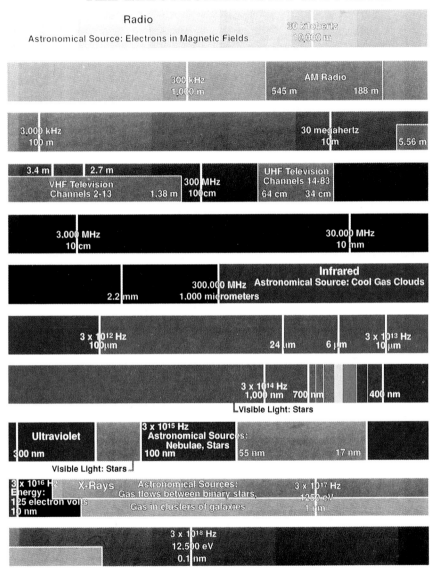

Figure 16.1. The electromagnetic spectrum.

The Cosmic Primordial Electromagnetic Radiation Field

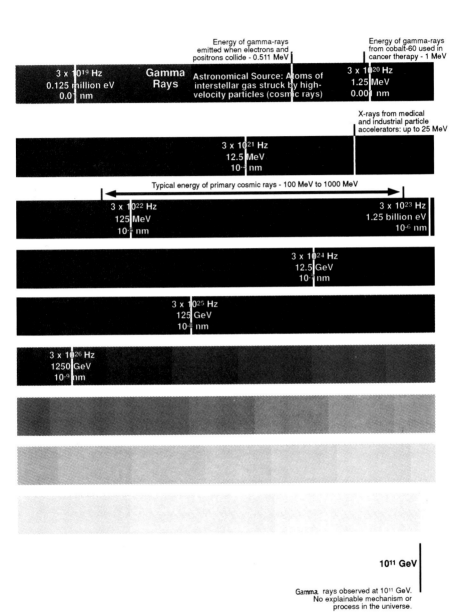

Figure 16.1. (*Continued*)

lengths and energy levels plays a vital role in our own universe and helps us to follow major events. Although we have ventured with automatic spacecrafts only beyond the planet Neptune, we are already able to discover astronomical objects and their activities at distances as far away as 15 billion light-years. At a distance of 15 billion light-years, we observe events when our universe was young and the early galaxies and quasars were forming and acting violently.

All the information and secrets of the universe are carried to us by vibrating electric and magnetic fields or electromagnetic radiation from which the visible light is the most common form. It is obvious to me that the electromagnetic radiating energy must be the basic element of the cosmos.

Everything in space, warmer than the absolute zero 0 K or $-237\,°C$, emits electromagnetic radiation in one form or another. By detecting the radiation and analyzing its frequency, wavelength, energy level, and changes caused by moving objects or the doppler effect, we can determine such properties of astronomical objects as density, temperature, chemical composition, distance, strength of magnetic fields, speed of movement, and other characteristics. The higher the frequency and smaller the wavelength, the higher the energy level of the radiation.

In addition to visible light, the spectrum of electromagnetic radiation, as shown in the graphics of Figure 16.1, covers long-wave, low-frequency radio waves at 30 kHz frequencies and 10,000 m wavelengths, reporting on electrons in magnetic fields of stars, to 10 cm wavelengths and 3,000 MHz frequencies, allowing radio-telescopic images of events in galaxies and intergalactic space. The cosmic background radiation peaks at 1 mm and 300,000 MHz. Infrared radiation appears at 300,000 MHz and 1000 μm to 3×10^{12} Hz and 100 μm revealing cool gas clouds in interstellar space.

The visible light which allows observation of the universe with telescopes comes in on frequencies from 3×10^{14} Hz, 1000 to 400 μm, where ultraviolet radiation starts and allows the detection of ultraviolet emitting stars. Only the visible, radio, and infrared signals can be detected on earth at ground level. All other shorter-wavelength, high-frequency radiation such as ultraviolet, x-ray, and γ-ray emitting sources can only be detected by special orbiting detectors from outer space beyond the earth's atmosphere, which either absorbs or deflects the shorter-wavelength electromagnetic radiation.

There seems to be no upper limit to the energy level of short-wavelength, super-high-frequency radiation. They extend from x rays at 3×10^{16} Hz, 10 m wavelength and energy levels of 125 eV to 3×10^{19} Hz 0.01 μm and energy levels of 0.125 million eV, revealing events taking place in gas flows between binary stars and events around black holes and quasars.

Even shorter and more powerful radiation is generated in the universe in the form of γ rays, starting with 10^{19} Hz, 0.125 million eV, and 0.01 μm emitted when atoms in interstellar space are struck by high-energy, high-velocity γ rays with wavelengths of anywhere from 10^{-6} μm to 3×10^{23} Hz frequency and energies of 1.25 GeV up. Gamma (γ) rays behave like energy bullets rather than waves. Bright sources of γ rays at lowr energy levels are pulsars or rapidly rotating neutron stars with powerful magnetic fields which then, near the poles, release synchotron radiation. Another powerful source of gamma rays is the infalling material from a binary star into the accretion disc of a black hole, originally the companion star.

It is clear from this short analysis of the electromagnetic spectrum that electromagnetic radiation is not only vital to our understanding of the universe but, together with the particles of matter and the four forces, form the basic elements of the cosmos. Matter can be annihilated and turned into energy in the form of γ rays, which is a routine occurrence in high-energy particle accelerator experiments. The three forces of nature lose their individual identity at energy levels above 10^{15} GeV and unify. Gravity joins the unification at levels above 10^{19} GeV, which prevailed at the birth of the universe. The radiation, particles, and four forces are all often called fields.

It is logical to conclude, therefore, that the "seeds" of gravity must also have been contained in the form of virtual gravitons in the primordial cosmic radiation. Once matter was created, they became real gravitons and mediated gravitation forces that affect matter and space.

Direct detection of this all-powerful cosmic, primordial radiation has not been possible as it flows only in the perfect vacuum of the "interuniverse" space of the cosmos. But it can reasonably be assumed that remnants of this radiation are contained in the present universe as γ rays and have already been discovered by orbiting observatories at enormous energies of up to 10^{11} GeV. (See Figure 16.1, right-hand page.) No known process in our universe, even in the most powerful quasars or cosmic black holes located in the centers of galaxies which swallow up to 100 million stars, can explain the origin of such high-energy fields. To generate γ rays at such energy levels would require temperatures of hundreds of billions of degrees and such temperatures do not exist even in quasars. They definitely cannot be of thermal origin and are possibly remnants of the cosmic primordial radiation trapped in the fireball during creation—a hypothesis which I introduced in 1985 in my theory of creation.

In September 1991, when I was making the final corrections to this manuscript, a NASA orbiter discovered powerful bursts of γ radiation which are isotropic over the skies and, therefore, indicate that they may be of cosmological origin. The bursts may indicate scattering by large masses of matter in the early universe, preventing the escape of this powerful radiation.

17

The Creation Process

17.1. PRESENT THEORY

The creation process of matter and radiation contained in the universe has intensively occupied the minds of cosmologists for over 30 years and has become a specially controversial subject. How does the "standard" singularity theory deal with particle creation? Any questions relating to the time before the singularity are considered meaningless, as there was no before, only nothingness. Time, matter, and radiation started with the big bang or rather 10^{-43} seconds after, when the size of the universe was only 10^{-33} cm. The entire universe at that time was a million, billion times smaller than a single electron (10^{-17} cm).

Many attempts were made to establish scenarios for particle creation. All of them, without exception, violate the laws of the conservation of energy. In the typical "free lunch" theory, particle creation is triggered by a vacuum fluctuation creating one pair of elementary particles and then spreading in an uncontrolled manner. In this theory, as in all other theories, the law of conservation of energy is violated. Experimental evidence clearly indicates that an outside source of concentrated energy must be introduced to create particles of matter from the vacuum of space. The virtual particles of the vacuum can become real particles only when a concentrated high-energy flash of γ rays is induced into the vacuum, equal in energy to at least the rest mass of the particles.

If the singularity could be considered a cosmic black hole, we could apply presently existing theories. According to Hawking, all black holes slowly lose particles which escape the black hole event horizon. Mass and energy are gradually lost this way and the rate of loss depends on the "size" of the back hole. According to calculations using Hawking's equations, it would take, for the galactic scale singularity, 10^{100} years to fully evaporate.

It is also my firm belief that the creation of matter and radiation contained in the universe must have been the result of events similar to those which take place in high-energy particle accelerators—a straight conversion of highly concentrated energy directly into particles of matter, having rest masses at least equal to the energy input. A direct transformation of energy into mass, following Einstein's equation of equivalence of mass and energy:

$$m = \frac{E}{c^2}$$

Nobel prize winner Steven Weinberg, in his book *The First Three Minutes*, expresses, indirectly, the idea that the singularity was filled with pure radiation energy and after the big bang explosion:

- "... the temperature of the radiation was so high that collisions of photons with each other could produce material particles out of pure energy."
- "In order for two photons to produce two material particles of mass in a head-on collision, the energy of each photon must be at least equal to the rest energy $E = mc^2$ of each particle."
- "Since the characteristic photon energy is the temperature times the Boltzmann's constant $E = Ta$, the temperature of the radiation has to be at least of the order of the rest energy mc^2 divided by the Boltzmann's constant, which must be reached before particles of this type can be created out of radiation energy."

And so each particle of matter depending on its rest mass has a threshhold temperature, which, for a pair of the smallest elementary particles, an electron e^- and positron e^+, is 6×10^9 K or 6 billion degrees Kelvin. Such a high temperature does not exist today anywhere in the universe and, therefore, no new electrons or positrons pop out of empty space in today's universe when light or even gamma radiation pass by.

Steven Hawking, in his best-seller *A Brief History of Time*, explains that "at the Big Bang itself, the universe is thought to have had zero size and so it must have been infinitely hot." He does not give any clue as to the content of the singularity before and at the time of explosion. As to the origin of particles of matter, he refers to the theory of inflation by Guth, described earlier in this book. Hawking, in his only reference to particle creation in the book, describes the theory as follows: "The idea of inflation could also explain why there is so much matter in the universe. There are something like 10^{80} particles in the region of the universe we can observe." Where did they all come from? The answer is that in quantum theory, particles can be created out of energy in the form of particle–antiparticle pairs. But that raises the question of where the energy came from. The answer is that the total energy of the universe is exactly zero. The matter in the universe is made up of positive energy, however, all the matter attracts itself by gravity. Thus, in a sense, the gravitational field has negative energy.

In the case of the universe that is approximately uniform in space, we can show that this negative gravitational energy exactly cancels out the positive energy represented by the matter. So the total energy of the universe is zero. Now, twice zero is also zero. Thus, the universe can double the amount of positive matter energy and also double the negative gravitational energy without violation of the conservation of energy.

This is what happened in the inflationary expansion of the universe, because the energy density of the supercooled state remains constant, while the universe expands. When the universe doubles in size, the positive matter energy and the negative gravitational energy both double, so the total energy remains zero. Thus the total energy during the large inflationary expansion phase, available to make particles, becomes very large. So here, Stephen Hawking believes in the "free lunch" theory of particle creation from the vacuum of space energy during the short inflationary expansion 10^{-35} seconds after the big bang, which supposedly lasted for 10^{-30} seconds. Later in the book Hawking says that "God may know how the universe began, but we cannot give any particular reason for thinking it began one way rather than another" and, even later, "the universe would be completely self-contained and not affected by anything outside itself. It would neither be created nor destroyed. It would just be."

The only other slightly different reference to particle creation that I have found in cosmological literature is the process described by Joseph Silk in *The Big Bang*. "Could not the intense gravitational field itself result in the creation of matter and radiation out of the vacuum? The very early universe might have been empty!" He continues, "The immense tidal gravitational forces that consequently resulted from the Big Bang can be imagined as disrupting the continuum of space-time in the process of creation. One can think of the vacuum as containing virtual pairs of particles and antiparticles. A sufficiently intense tidal gravitational field can disrupt these virtual pairs, releasing the particles into the real world."

I assume that the author believes in a cycling universe when, during the implosion into a singularity or cosmic black hole, the entire mass of particles and radiation of the collapsing universe was transformed at high-density levels into pure gravitational energy, equal to the rest mass of matter in the universe, and that this gravitational energy exploded during the big bang creating time, space, and particles. Again, it is obvious that this process could not apply or explain the creation of the universe of matter and radiation for the first time.

17.2. THE VELAN THEORY OF CREATION

I can now finally describe, for the first time, the complete sequence of events that brought about the birth of our universe, the act of creation itself.

As is obvious, my theory is substantially different from anything else written on this subject. My theory eliminates the mysterious singularity and has a true beginning. The revolutionary hypothesis should not be considered a disadvantage, as significant progress is impossible in a situation of self-satisfying unanimity. I will now take you on a journey to the edge of space, time, and comprehension.

The "beginning" in my theory of creation is extremely simple and harmonious. It is in full compliance with the physical sciences and laws of nature. It does not violate the conservation laws and gives a full account of the origin of all matter, radiation, and the four forces, as well as the seeds of growth and evolution setting off a chain of complicated processes and transforming simplicity to the present state of our universe.

The theory is a combination between extrapolation of proven physics and observational data and a basic, new and powerful hypothesis from which many fruitful deductions can be made, explaining what has so far remained inexplicable. I think that the value of my hypothesis must be judged not from the standpoint of how plausible it sounds but, rather, by the self-consistent scheme formed as a consequence of the hypothesis and which agrees with actual observations backed up by laboratory experiments.

Any viable cosmological theory must account for the origin of all matter and radiation contained in the universe.

The creation process contains five distinct events. What follows is a recap of the creation scenario, which will be described in more detail in the following chapters.

17.3. A SHORT PREVIEW OF THE VELAN THEORY OF CREATION

1. *The creation moment.* A cloud of elementary particles of matter and electromagnetic radiation is created from a vacuum of space interacting with the primordial cosmic radiation fields flowing through the cosmos. A combination of violent fluctuations in the superspace cells (10^{-33} cm) and intensive appearance of virtual particles establishes the base for interaction.

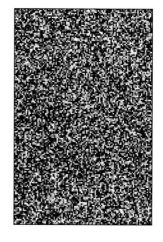

2. *The "primordial fireball"*. The just-created cloud of matter and radiation is transformed by gravitational forces which act as soon as matter appeared in the primordial fireball of hot elementary particles of matter and radiation.

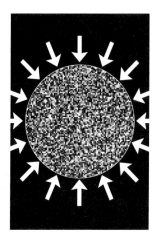

3. *Gravitational collapse–mechanical equilibrium*. Under gravitational forces, the fireball collapses further and particles rush toward the center at velocities near the speed of light. Their kinetic energy is transformed into heat and the temperature of the fireball rapidly increases to intense levels. As gravitational forces are the largest in the center, a highly dense core is created resisting the squeeze. For a short moment the gravitational forces equal the thermal forces of particles and radiation acting outwards.

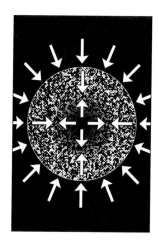

4. *Gravitational collapse continues–core achieves enormous density.* The gravitational forces trying to collapse the fireball into a black hole squeezed the core to densities far beyond the quark–electron density. The core bounces back and creates shock waves.

5. *The big bang of the fireball.* The shock waves created by the rebouncing of the squeezed core, together with the overwhelming thermal energy of radiation trapped in the fireball, overcame the gravitational implosion and the fireball erupted in a titanic explosion called the big bang, very similar to supernova explosions of large dead stars. The core collapsed into a black hole.

The Creation Process

We can now look the moment of creation directly in the eye and follow the scenario in more detail. It is important to accompany the nonmathematical discussions and presentations with simple, clear mathematics in order to substantiate the new theory, using proven thermodynamic and gravitational equations. The early universe contained only three particles of matter—electrons and u and d quarks. The same particles exist in the universe today, except that the then-free quarks are now all confined in protons and neutrons.

The important property of a particle that relates to the creation process is the *rest energy* $E = mc^2$ or the energy that would be released if all of the mass m of the particle were transferred into energy. The other property of interest is the *threshold temperature* T_t, which is the rest mass or energy of the particle divided by the Boltzmann constant $k = 1.38 \times 10^{-16}$ erg/K. As the rest mass is usually presented in astrophysics in millions (MeV) or billions (GeV) of electron volts we must convert k from ergs to electron volts (1 erg = 0.63×10^{12} eV):

$$k = 0.00008617 \text{ eV/K}$$

The threshold temperature is the level above which a particle can be created out of radiation and can be calculated from the following equation:

$$T_t = \frac{mc^2}{k} \quad \text{(K)} \qquad (16.1)$$

For a photon with zero mass, $T = 0$. Properties calculated from Eq. (16.1) are shown in Table 17.1.

It can be seen that quarks are created above a temperature of 3.692×10^{12} K and electron–positron pairs above 11.86×10^9 K. The threshold temperature of 5.93×10^9 K is the minimum level for creation of electrons.

Table 17.1. Threshold Temperatures According to Eq. (16.1)

Particle	Symbol	Rest energy mc^2 (MeV)	Threshold temperature ($\times 10^9$ K)	Lifetime (sec)
Photon	γ	0	0	Stable
Neutrinos	$\nu_e, \bar{\nu}_e$ $\nu_\mu, \bar{\nu}_\mu$ $\nu_\tau, \bar{\nu}_\tau$	0 (possible 10–30 eV)	0	Stable
u quark	u, $\bar{u}$	312.75	3692	Stable[a]
d quark	d, $\bar{d}$	312.75	3692	Stable
Electron	e^-, e^+	0.511	5.93	Stable

[a] Can decay into a d quark by the beta decay of a neutron.

At the time of creation, the prevailing temperatures were as high as 10^{26} K and for this reason particles were created in extremely large numbers in the presence of an outside source of radiation. In addition, when the temperature is higher than the threshold level, the particles created acquire higher energy and collide even more violently with each other.

It is remarkable to note that even in the largest stars, the highest operating temperatures in core burning of silicon into iron reach 3.5×10^9 K (recall that 6×10^9 K is required to create an electron). At a nuclear density of 3×10^{14} g/cm^3, a neutron star may reach a temperature of 100×10^9 K (the threshold temperature for creating a quark is 3.7×10^{12} K). Such high temperatures are not achieved in any processes in our universe, not even in or around quasars or cosmic black holes of up to 8 billion solar masses.

17.4. THE CREATION MOMENT

On rare occasions, cosmically speaking, two vital elements instrumental in the particle creation process interact in a unique and optimum manner: the primordial, intense radiation field and the vacuum of the interuniverse cosmic space, which simultaneously experiences large fluctuations in its topology and virtual particle activity, as described in the postulates to the theory.

Suddenly, whether by pure chance or determined and directed by divine power, a secret of nature that may remain forever a mystery to intelligent beings, the particle creation process, is triggered (Figure 17.1; see page 201).

The vacuum of space in a given area of the cosmos violently fluctuates. The tiny 10^{-32} cm cells of space undergo unusually violent dynamic changes in their geometry. Space vibrates, expands, and explodes. At the same time there is an unusual activity and massive appearance of virtual particles, ready to be transformed into particles of matter. Due to the high density (10^{94} g/cm^3) of virtual particles in the vacuum of space, the potential for creation in the presence of a powerful energy source is enormous. The unusual activity of virtual particles may have been triggered by the intense vacuum fluctuations of the topography and geometry of space. The intensive fluctuations of the superspace created a sort of tidal wave, disrupting the virtual particles. When conditions reached critical levels, the omnipresent, primordial cosmic radiation flowing through the interuniverse space at velocities greater than the speed of light was suddenly slowed down and compressed by the unusual fluctuations in the geometry of space. Its energy level intensified to creative levels of 10^{12}, 10^{13} GeV and what followed was a massive transformation and release of the virtual particles into the real world. Pairs of quarks–antiquarks and electrons–positrons were created in an

THE CREATION MOMENT

LOCATION: VACUUM OF SPACE IN THE MULTIUNIVERSE COSMOS.
PROCESS: VIOLENT FLUCTUATION OF SUPERSPACE IN TINY CELLS OF 10^{-33}cm COMPRESS THE INFLOWING PRIMORDIAL COSMIC RADIATION OF 10^{12} GeV, INCREASING ITS STRENGTH. SIMULTANEOUS ENORMOUS VACUUM FLUCTUATIONS OF VIRTUAL PARTICLES, WHICH BECOME REAL PARTICLES, THEIR MASS SUPPLIED BY THE ENERGY OF THE PRIMORDIAL RADIATION FIELD.

instantaneous gigantic direct transformation of electromagnetic energy into matter, following the equivalence formula of Einstein $m = E/c^2$. The powerful primordial radiation energy gave the virtual particles their mass.

A new universe was created. The task was carried out jointly by the basic energy field of the cosmos together with the cosmic vacuum of space, which both carry in them the seeds of matter of the four forces of nature and all of the elements of future development, unfolding from unique simplicity to a high level of complexity.

This is the event that took place approximately 18 billion years ago and that created in a split second a cloud of matter and radiation.

Some of the primordial radiation speeding away became trapped with the particles of matter and accounts to a large extent, for the overwhelming disproportion of photons versus particles of matter in the fireball.

There are 10^9 photons in our universe for each baryon (proton or neutron). The same relationship prevailed in the early universe and the only viable, scientific explanation for the large content of radiation is the primordial cosmic radiation.

The photons left from the annihilation process when pairs of electrons–positrons and quarks–antiquarks collided and created photons, cannot alone account for the disproportion as many of the annihilation photons collided in pairs and in turn created new particles. Also, the discovery of γ rays in our universe at the enormous energy level of 10^{11} GeV supports the theory. The recent NASA γ orbiter detected powerful bursts of γ rays, confirming earlier discoveries.

As there is no mechanism or process taking place in the cores of galaxies, stars, and quasars that could account for the presence of such powerful electromagnetic radiation, it is reasonable to assume that the detected γ rays are remnants of the primordial cosmic radiation field trapped in the fireball during the creation process. Hopefully, this and other discoveries in the future will prove that the primordial cosmic radiation theory introduced for the first time in my cosmological model is a sound contribution toward elucidating the mystery of creation.

The particles born from the conversion of energy into matter during this upheaval of creation consisted of pairs of electrons–positrons, quarks–antiquarks, neutrinos, and photons. All three forces of nature were unified and did not interact with particles at the high energy and temperature levels. Only gravity acted with increased intensity as soon as particles of matter appeared.

Within about 10^{-23} sec, annihilation of particles and antiparticles (positrons, antiquarks) took place creating as a result more electromagnetic energy in the form of γ rays and, in turn, more particles. Ultimately, only matter remained and the "fireball" plasma soup consisted of electrons, quarks, photons, and neutrinos.

The particle creation process stopped when the fluctuations of the vacuum of space diminished to normal levels.

17.5. WHY MATTER ONLY?

There is no evidence in the universe for the presence of antimatter in the form of particles or galaxies. Except in collisions of cosmic rays in the upper atmosphere and high-energy accelerators where antiprotons or positrons are created, the whole universe, as we know it, consists of matter.

The prevailing temperature in the just-created particle–radiation cloud was extremely high (10^{15}–10^{20} K) and allowed elementary particles to transform quarks into positrons, antiquarks into electrons, and electrons into quarks and antiquarks. The main reason why, at the end of this short period lasting approximately 10^{-23} to 10^{-10} sec, we were left only with particles of matter and not antimatter is the fact that obviously the laws of physics at superhigh levels of energy and temperature did not apply to the same extent or equally to particles and antiparticles. Nature has a definite preference for matter.

Normally, under energy and temperature levels prevailing in the present universe, nature follows symmetry. There are three basic symmetries—C, P, and T. The C symmetry specifies that the laws of physics are the same for particles and antiparticles. Symmetry P deals with the mirror image of particles, right- and left-spinning electrons, for instance. Finally, the T symmetry tells us that the laws are the same in the forward and backward direction of time. There are, however, exceptions to the CPT symmetries even in the present universe. Electrons and the weak nuclear interactions, for instance, do not obey the CP symmetry. We must assume that the CPT symmetries did not apply at the high energy and temperature levels that prevailed in the hot primordial soup and, consequently, more positrons turned into quarks and antiquarks into electrons than vice versa. The excess of quarks and electrons made up the matter, as we see it today in the universe.

During that short period, neutrinos (ν) were produced as intensively as photons (γ). Typical interactions are shown in Figure 17.2.

After the short period of creation and transformation, which lasted approximately 10^{-10} sec, the total estimated mass of particles and photons exceeded approximately 25% of the critical mass in the universe today. The critical mass is the level required to stop the expansion of the universe. A slight excess would be sufficient to reverse the expansion and initiate a collapsing implosion.

A large portion of the excess mass was contained in the core of the fast-forming fireball, and the core, exposed to intense gravitation, collapsed finally into a black hole.

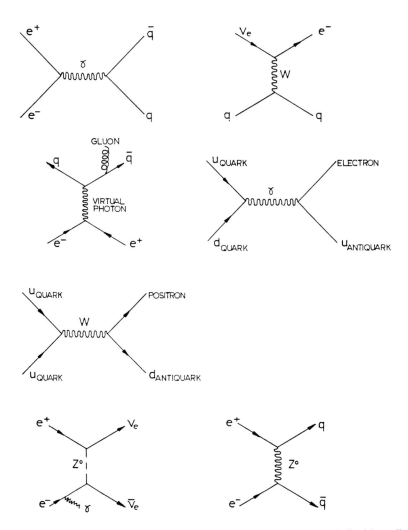

Figure 17.2. Interactions of leptons (electron, positron, neutrino) and quarks u, d. Particles called mediators mediate the interactions. If two particles pass close to each other or collide, one of them will emit a mediate (γ, $W^\pm$, Z^0, gluons, or gravitons) and the other will absorb it. If the particles are sufficiently energetic, one of them, say an electron, can emit a Z^0 and then the positron will annihilate with the electron, leaving the Z^0 free for a moment. Soon after, the Z^0 will decay back into a pair of elementary particles, electron–positron or quark–antiquark.

The Creation Process

At the end of the creation period of matter and radiation, the fireball contained:

$$3 \times 10^{80} \text{ quarks u and d}$$
$$10^{80} \text{ electrons}$$
$$10^{89} \text{ photons}$$
$$10^{89} \text{ neutrinos}$$

Today, the universe contains an estimated 10^{80} protons and neutrons, formed later at lower temperatures from the 3×10^{80} quarks, 10^{80} electrons, 10^{89} photons, and 10^{89} neutrinos. In essence, the present relationship between particles and matter in the universe is the same as in early times.

18

The Velan Fireball

Less than a second after the creation of the primordial cloud of particles of matter and radiation, intense gravitational forces appear simultaneously with the appearance of matter and collapse the cloud into a fireball that is subjected to continuous and rapid implosion. The kinetic energy ($Mv^2/2$) of the infalling particles at velocities close to the speed of light and the powerful collisions between particles and scattering of photons all turn into heat, energy, and pressure, counteracting the gravitational forces. (See Figure 18.1, page 208.)

Radiation and the particles of matter are in full thermal equilibrium—particles are equally as hot as photons. The photons cannot escape. They move with the speed of light between collisions with electrons and quarks, are scattered and contribute to the largest extent to the ever-increasing temperature in the fireball. The energy of photons increases with the fourth power of temperature. The energy level is approximately 10^{13} GeV, temperature 10^{20} K, matter density 3.7×10^8 g/cm^3, and the radius of the fireball 19×10^{15} cm. Implosion continues at a rapid pace.

At the center, due to more intensive gravitation, a denser core of particles and radiation is being formed. Events are depicted in Figure 18.2 (see page 209).

Gravitational energy, instrumental in imploding the newly created fireball, is counteracted by thermal forces of matter and radiation (see Figure 18.3). We must now analyze these forces and determine their interactions.

18.1. GRAVITATION E_{GRAV}

The gravitational forces started to act on the cloud of matter and radiation as soon as they appeared. The cloud was instantaneously transformed into a regular fireball, and the gravitational field, generated not only by the particle mass but

GRAVITATIONAL COLLAPSE OF THE JUST CREATED "FIREBALL" OF SUPERHEATED PRIMORDIAL SOUP OF PARTICLES AND RADIATION.

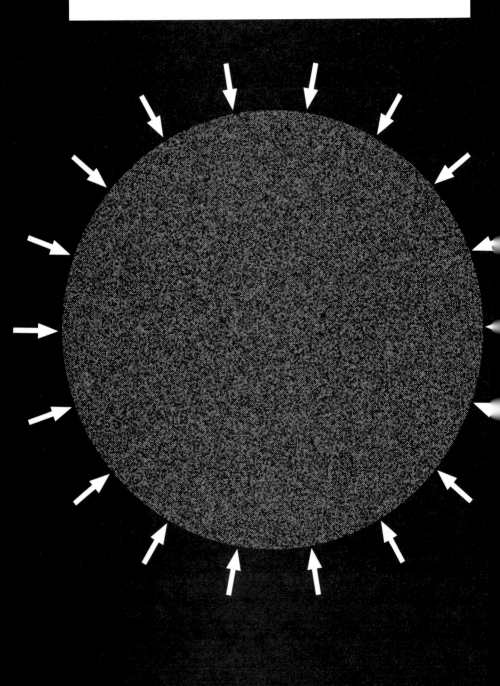

IMPLOSION CONTINUOUS IN RAPID PACE.
A DENSER INSIDE SPHERE OF PARTICLES AND RADIATION
IS BEING CREATED DUE TO MORE INTENSIVE
GRAVITATION CLOSE TO CENTER.

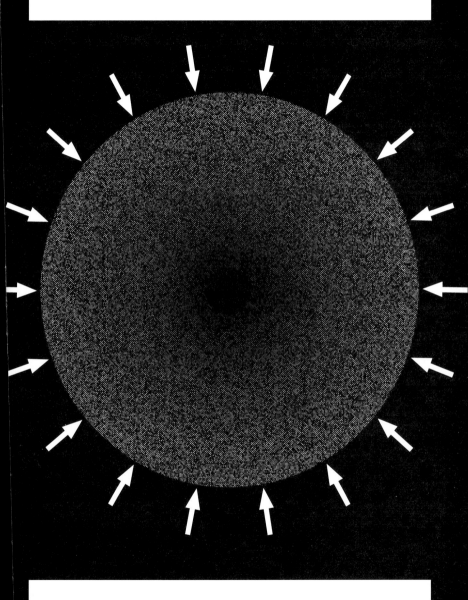

FIRST PHASE

RADIUS OF FIREBALL: 19×10^{15} cm, 190 billion Km
MATTER DENSITY: 3.7×10^8 g / cm^3
AVERAGE TEMPERATURE: $\sim 10^{20}$° K
ENERGY LEVEL: $\sim 10^{13}$ GeV

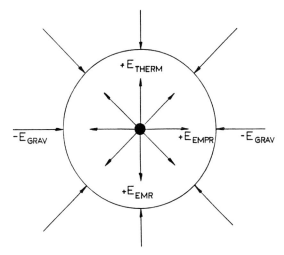

Figure 18.3. The four forces acting in the fireball: the gravitation, the energy of the thermal motion of particles, the energy of the electromagnetic radiation (photons), and the electromagnetic radiation pressure.

also by all forms of energy including photons, initiated the implosion squeeze with the ultimate aim to create a black hole or singularity. This never materialized.

In my theory, the highest level of energy achieved just before the explosion of the fireball is 10^{18} GeV, with an average temperature of 3.95×10^{26} K. The gravitational force $-E_{\text{GRAV}}$ acting toward the center $(-)$ of the fireball can be calculated from the following formula:

$$-E_{\text{GRAV}} \cong \frac{GM^2}{R_{\text{FB}}} \quad (\text{g-cm}^2/\text{sec}^2) \tag{18.1}$$

where M is the mass of the universe, G is the gravitational constant, and R_{FB} is the radius of the fireball.

18.2. ENERGY OF ELECTROMAGNETIC RADIATION (PHOTONS) E_{EMR}

The electromagnetic radiation in the form of photons played the predominant role in counteracting the gravitational implosion effect. The energy of the photons rising with the fourth power of the temperature prevented the collapse of the universe into a black hole. The energy of photons E_{EMR} can be calculated from the following equation:

$$+E_{EMR} = aT^4 \quad (\text{erg/cm}^3) \qquad (18.2)$$

where T is the temperature in K and a is the radiation constant = 7.5647×10^{-15} erg/cm^3 K^4.

18.3. ENERGY OF ELECTROMAGNETIC RADIATION PRESSURE E_{EMRP}

The thermal energy of the photons also creates an outward pressure that amounts to one-third of the electromagnetic radiation energy:

$$E_{EMRP} = \tfrac{1}{3} E_{EMR} = \tfrac{1}{3} aT^4 \quad (\text{erg/cm}^3) \qquad (18.3)$$

18.4. ENERGY OF THERMAL MOTION OF PARTICLES E_{THERM}

The particles in the fireball fall with enormous speed toward the center under the squeeze of gravity forces but also collide continuously with each other and the photons, contributing to the increase of the overall temperature which is equal for the particles and photons, with complete thermal equilibrium. Some scientists go as far as to consider their behavior and energy contribution to be equal to that of photons and would calculate their energy from the formula $E = aT^4$ (erg/cm^3).

I have chosen to calculate the thermal energy contribution of the particles from the thermodynamic formula for particles of matter in motion at temperature T (K):

$$E_{THERM} = 3(N)kT \quad (\text{erg/K} \times \text{K} = \text{erg}) \qquad (18.4)$$

where N is the number of particles and k is the Boltzmann constant = 1.38×10^{-16} erg/K. The forces resulting from thermal energy, acting outwards against the negative force of gravitation, are shown in graphical form in Figure 18.3 and below:

$$+E_{THERM} \rightrightarrows + E_{EMR} \rightrightarrows + E_{EMRP} \rightrightarrows \overset{+++}{\longrightarrow} \quad \overset{-}{\longleftarrow} -E_{GRAV}$$

These formulas allow us to calculate at any time of the implosion process the radius, density, temperature, and imbalance of all acting forces.

I would like to emphasize, especially for those who have read Steven Weinberg's *The First Three Minutes*, that the fireball did not include any protons or neutrons. Their constituent quarks u and d are free particles together with electrons, photons, and neutrinos. The quarks are close to each other, enjoy their "asymptotic freedom," and the strong atomic force is inactive at this high level of temperature. At a later time, after the explosion of the fireball when temperatures drop to 10^{12} K, quarks will lose their freedom and find refuge in protons and neutrons.

18.5. EQUILIBRIUM

The implosion continues at a rapid pace. At a certain point and for a very short period of time, maybe 10^{-10} sec, the gravitational forces are in equilibrium with the thermal forces (see Figure 18.4, page 213).

We can calculate the radius of the fireball R_{FB} at the equilibrium stage from the well-known virial theorem used in astronomy. Some will say that a system collapsing at those enormous velocities will not satisfy the virial theorem applying to stars. However, I feel that it is a good basis for approximation.

According to this theorem, for the fireball to be in mechanical equilibrium, E_{THERM} and E_{GRAV} are related by the expression:

$$2E_{THERM} + E_{GRAV} = 0$$

where

$$E_{THERM} = 3(N)kT$$

$$N = \text{number of particles} = \frac{M(\text{total mass})}{m_{\text{particles}}}$$

$$3\frac{M}{m}kT = \frac{GM^2}{R_{FB}}$$

$$R_{FB} = \frac{GM}{3kT} \times m_{\text{particles}}$$

where M is the total mass of the fireball including matter, photons, and neutrinos. We are taking the critical mass + 25% (5.7 × 10^{56} g + 25%) as at least 20% of the mass remained in the core after explosion. The core collapsed into a black hole.

$M = 7.1 \times 10^{56}$ g

G = gravitational constant 6.673×10^{-8} cm^3/g sec^2

k = Boltzmann's constant 1.38×10^{-16} erg/K

IMPLOSION CONTINUOUS AT A RAPID PACE.
FIREBALL IS SHORTLIVED IN MECHANICAL EQUILIBRIUM.

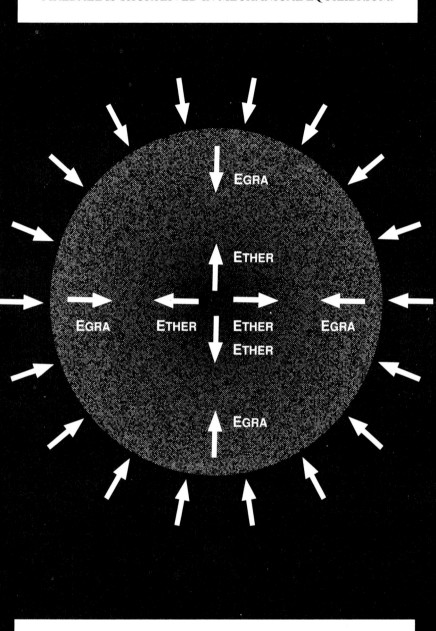

EQUILIBRIUM PHASE
RADIUS OF FIREBALL: 9.59 x 10^{15} cm, 95.9 billion Km
AVERAGE TEMPERATURE: ~$10^{21°}$ K
ENERGY LEVEL: ~10^{14} GeV

T = temperature of the fireball, estimated at 10^{25} K

m = average mass of particles in grams, calculated as follows:

mass of electron $m_e = 9.109 \times 10^{-28}$ g

mass of proton $m_p = 1.672 \times 10^{-24}$ g

mass of quark $\dfrac{m_p}{3} = 0.55 \times 10^{-24}$ g

average mass $m = 0.838 \times 10^{-24}$ g

The radius of the fireball at equilibrium R_{FB} is calculated as follows:

$$R_{FB} \cong \dfrac{6.673 \times 10^{-8} \times 7.1 \times 10^{56}}{3 \times 1.38 \times 10^{-16} \times 10^{25}} \times 0.838 \times 10^{-24}$$

$$\cong \dfrac{39.7 \times 10^{24}}{4.14 \times 10^{9}}$$

$$\cong 9.59 \times 10^{15} \text{ cm}$$

The universe at the time of equilibrium had a radius of 9.59×10^{15} cm. For comparison, the radius of the sun, which is a small star, is 7×10^{10} cm.

18.6. FIREBALL COLLAPSES TO QUARK–ELECTRON DENSITY

Shortly after the equilibrium phase the fireball reached the quark–electron density and the moment of explosion was nearing (see Figure 18.5, page 215). What was the radius of the fireball at that moment?

The fireball contained (present content of the universe + 25%):

4.395×10^{80} electrons

13.185×10^{80} quarks

4.395×10^{89} photons

4.395×10^{89} neutrinos

Total minimum volume taken up by all matter and radiation when all particles were squeezed to quark–electron density was:

Electrons: Volume $= \pi \times (10^{-16})^3 \times 4.395 \times 10^{80}$

$= 13.80 \times 10^{32}$ cm^3

Quarks, u, d: Volume $= \pi \times (0.5 \times 10^{-13})^3 \times 13.185 \times 10^{80}$

$= 5.14 \times 10^{41}$ cm^3

Photons: Number $N = 20.3 \times T^3$/cm^3

IMPLOSION OR THE SQUEEZING OF THE FIREBALL CONTINUES. THE DENSITY OF THE CENTER SPHERE ACHIEVES QUARK-ELECTRON OR EQUILIBRIUM DENSITY.

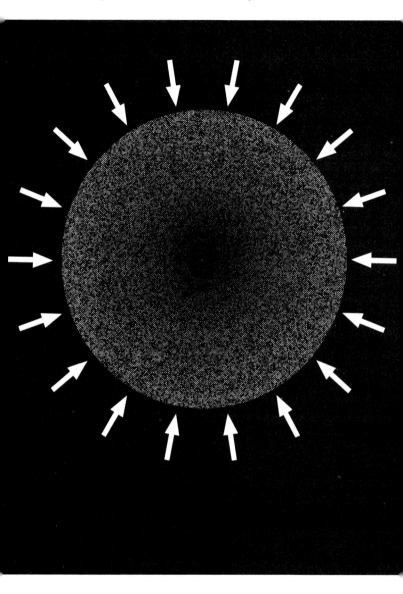

SECOND PHASE
RADIUS OF FIREBALL: About 1.17×10^{14} cm, 1.17 billion Km
DENSITY OF CORE: 1.3×10^{15} g cm^3
DENSITY OF FIREBALL: CLOSE TO QUARK-ELECTRON DENSITY
AVERAGE TEMPERATURE: 3.95×10^{26} °K
ENERGY LEVEL: 10^{18} GeV

The average temperature of the fireball, calculated later, at this time was $T = 10^{26}$ K:

$$N = 20.3 \times (10^{26})^3 = 20.3 \times 10^{78} \text{ photons/cm}^3$$

$$\text{Total volume } V_{\text{ph}} = \frac{4.395 \times 10^{89}}{20.3 \times 10^{78}} = 0.22 \times 10^{11} \text{ cm}^3$$

The total volume of the fireball at quark–electron density was:

$$V_T = V_e + V_q + V_{\text{ph}} + V_r$$
$$V_T \cong 5.14 \times 10^{41} \text{ cm}^3$$

The radius of the fireball at quark–electron density was:

$$\pi R_{\text{QED}}^3 \cong 5.14 \times 10^{41}$$

$$R_{\text{QED}} \cong \left(\frac{5.14}{3.14} \times 10^{41}\right)^{1/3} \cong 1.17 \times 10^{14} \text{ cm}$$

The density was equivalent to $d = M_T/V_T$ or

$$d = \frac{7.1 \times 10^{56}}{5.14 \times 10^{41}} = 1.3 \times 10^{15} \text{ g/cm}^3$$

This figure compares with the nuclear density of neutron stars of 1.17×10^{14} g/cm^3.

The universe, which today extends over a space with a radius of 18 billion light-years, was compressed to 1.17×10^{14} cm or less than 1/10,000 of a light-year (1 light-year = 0.946×10^{18} cm).

The average temperature of the fireball at quark–electron density can be calculated as follows. The thermodynamic formula for the energy of thermal motion E_{THERM} is Eq. (18.4):

$$E_{\text{THERM}} = 3(N)kT$$

The formula for the gravitational energy E_{GRAV} is Eq. (18.1):

$$E_{\text{GRAV}} \cong \frac{GM^2}{R_{\text{FB}}}$$

where $G = 6.673 \times 10^{-8}$ cm^3/g sec^2, $M = 5.68 \times 10^{56}$ g + 25%, and $R_{\text{FB}} = 1.17 \times 10^{14}$ cm

$$3(N)kT = \frac{GM^2}{R_{\text{FB}}}$$

$$T = \frac{GM^2}{R_{\text{FB}} \, 3(N)k} \tag{18.5}$$

$$= \frac{6.673 \times 10^{-8} \times 50.41 \times 10^{112}}{1.17 \times 10^{14} \times 3 \times 17.58 \times 10^{80} \times 1.38 \times 10^{-16}}$$

$$= \frac{336 \times 10^{104}}{85.15 \times 10^{78}} = 3.95 \times 10^{26} \quad K$$

This is the average temperature in the fireball at the time the particles and radiation were compressed by gravity to quark–electron density (1.3×10^{15} g/cm^3). The temperature of the core by this time must have been considerably higher, reaching 10^{28}–10^{30} K.

19

The Big Bang of the Fireball

By the time the entire fireball reached the quark–electron density, the core was compressed far beyond this density level. All of the electrons, quarks, and other particles in the core merged to form a sort of single gigantic nucleus. A spoonful of such matter has the same mass as all of the buildings in Montreal combined. In this form, particles of matter show a powerful resistance to further compression. This, however, does not stop the outside layers of particles in the fireball from imploding further and exerting more squeezing power on the core.

At the surface of the hard core, the particles stopped suddenly, but not fully. The compressibility of elementary particles is low at nuclear density but not zero. The momentum of the infalling particles (mass m × velocity v), being close to the speed of light, compressed the central sphere to perhaps a density 4–5 times that of equilibrium, which we can call the point of the "maximum squeeze." By that time the thermal energy of particles and radiation, as calculated, reached a level higher than the total gravitational energy and exerted considerable pressure against the gravitational forces. While the internal pressure in the center of the sun's core is 10^{10} kg/cm^2, the pressure in the interior of the fireball was about 10^{31} kg/cm^2.

The core, after the "maximum squeeze," bounced back like a rubber ball that was compressed. The bounce set off enormous shock waves, which, together with the overpowering internal forces created mainly by the energy of the electromagnetic radiation, resulted in a titanic cosmic explosion, as shown schematically in Figures 19.1 and 19.2 (pages 200 and 221, respectively).

This theory is the first comprehensive, step-by-step description of the creation of matter and radiation contained in our universe, gravitation's attempt to collapse the newly born fireball, and finally the victory of the thermal forces of the superheated radiation which caused the ultimate explosion and set off the universe on its way to expansion, its creative history and glory of the last 18 billion years.

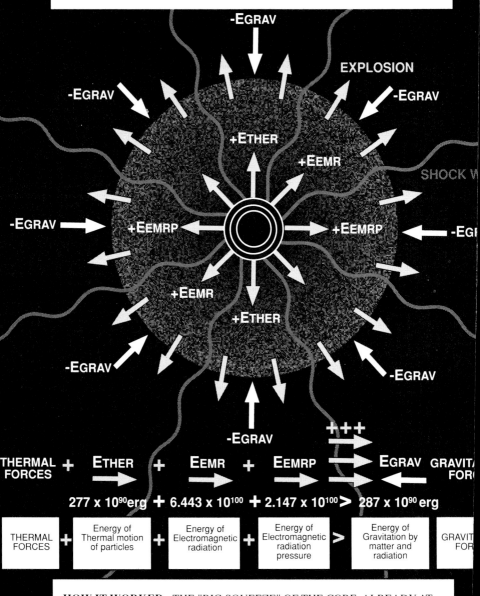

EXPLOSION PROCESS

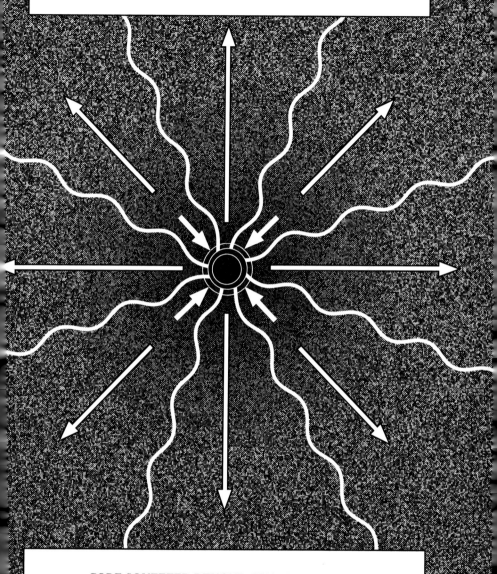

CORE SQUEEZED BEYOND QUARK-ELECTRON EQUILIBRIUM TO MAXIMUM SQUEEZE, BOUNCES BACK AND SETS UP ENORMOUS SHOCKWAVES. THIS TOGETHER WITH THE OVERWHELMING INTERNAL THERMAL ENERGY AND PRESSURE OVERPOWER THE GRAVITATIONAL FORCES RESULTING IN A COSMIC EXPLOSION SENDING THE UNIVERSE ON ITS NEAR TO SPEED OF LIGHT EXPANSION. THE CORE COLLAPSES INTO A BLACK HOLE.

We can now proceed to calculate all of the forces in the fireball at the time of the big bang. The large thermal forces of the radiation and the shock waves, which gave the explosion impulse, caused the cosmic explosion of the enormous mass of matter and radiation and set the universe on its expansion and development. We are using equations and data listed in previous chapters.

The force of gravitation E_{GRAV} is

$$E_{GRAV} = \frac{GM^2}{R_{FB}} = \frac{6.673 \times 10^{-8} \times 50.4 \times 10^{112}}{1.17 \times 10^{14}} = 287 \times 10^{90} \text{ ergs}$$

The thermal force of particles E_{THERM} is:

$$E_{THERM} = 3(N) \, kT$$

T has been calculated to be 3.95×10^{26} K

$$E_{THERM} = 3 \times 16.95 \times 10^{80} \times 1.38 \times 10^{-16} \times 3.95 \times 10^{26}$$
$$= 277 \times 10^{90} \text{ ergs} \quad (<E_{GRAV} = 287 \times 10^{90} \text{ ergs})$$

The result of the calculations indicates that the thermal energy of particles of matter would not overcome the gravitational forces and the universe would have collapsed into a black hole.

Energy of Electromagnetic Radiation E_{EMR}

The thermal energy of photons in the fireball was dominant, first of all, owing to the overwhelming quantity of photons originating mainly from the primordial cosmic radiation (4×10^9 photons versus 1 electron and 3 quarks) and its high energy level rising to the fourth power of the temperature, while the thermal energy of particles of matter depends largely on their rest mass.

We can calculate the energy level E_{EMR} per cubic centimeter from:

$$E_{EMR} = aT^4$$

where $a = 7.5647 \times 10^{-15}$ erg/cm^3 K^4 and the temperature just before explosion was $T = 3.95 \times 10^{26}$ K

$$E_{EMR} = 7.5647 \times 10^{-15} \times (3.95 \times 10^{26})^4$$
$$= 7.5647 \times 10^{-15} \times 243 \times 10^{104}$$
$$= 18.41 \times 10^{91} \text{ ergs/cm}^3$$

In order to determine the total energy, we must multiply E_{EMR} by the total volume of photons. The number of photons N per cm^3 can be determined from the equation:

$$N = 20.3 \times T^3 = 20.3 \times (3.95 \times 10^{26})^3$$
$$= 12.51 \times 10^{80} \text{ photons/cm}^3$$

The number of photons in the fireball was close to the number estimated in the universe today:

$$N_{TOTAL} = 4.395 \times 10^{89}$$

$$\text{Volume } V_{EMR} = \frac{N_{TOTAL}}{N} = \frac{4.395 \times 10^{89}}{12.51 \times 10^{80}} = 0.35 \times 10^9 \text{ cm}^3$$

$$= 3.5 \times 10^8 \text{ cm}^3$$

The total energy of the electromagnetic energy contained in the fireball is: $E_{EMR} \times V_{EMR}$ or

$$E_{EMRT} = 18.41 \times 10^{91} \times 3.5 \times 10^8 = 6.443 \times 10^{100} \text{ ergs}$$

$$\longrightarrow \quad \longleftarrow$$

$$\boxed{E_{EMRT} = 6.443 \times 10^{100} \text{ ergs} > E_{GRAV} = 287 \times 10^{90} \text{ ergs}}$$

The energy of the radiation pressure is

$$E_{EMRP} = \tfrac{1}{3}E_{EMR} = 2.147 \times 10^{100}$$

As can be seen, the outward forces acting against gravitation in the form of the thermal energy and pressure of photons were considerably larger.

The calculations are rather conservative. It is possible that the particles of matter, the electrons and quarks, which according to quantum mechanics can be looked upon as pointlike particles or as wavelike particles, acted in a similar way at these high temperature levels as photons. In this case the forces were even higher and the fireball would have exploded earlier. It really makes no significant difference nor does it digress from the main thrust of the theory.

The Velan story of creation is complete. The origin of particles and the large quantity of photons are well substantiated and the mysteries connected with the classical singularity–big bang theory are eliminated. All processes that took place comply fully with laboratory-proven physical laws of high-energy elementary particle physics and the laws of thermodynamics. While the size of the universe in the classical theory described in the book *The First Three Minutes* by Steven Weinberg was only 10^{-35} cm (smaller than a single electron, 10^{-17} cm), 10^{-45} sec after the big bang the Velan universe before explosion already had a sizable radius of 1.17×10^{14} cm. The radius of our sun, in comparison, is 7×10^{10} cm.

While there is no scientific explanation in the standard cosmological model of how matter was created or why the singularity exploded, the events in the Velan cosmological model are well substantiated. The explosion process is backed up mathematically, using standard and proven thermodynamic and astronomical equations.

20

The History of Evolution of the Early Universe

Before describing the step-by-step evolution of the universe after the explosion of the fireball, I would like to draw attention to the substantial difference between my theory and the classical singularity–big bang theory.

As shown in Table 20.1 and Figure 20.1, the classical theory does not give a clue as to how particles were created and even less information is available about the origin of the enormous volume of radiation contained in the universe. The history starts at 10^{-45} sec after the big bang when the entire universe had a radius of 10^{-50} cm, at a temperature of 10^{32} K and the energy level was at 10^{19} GeV. No known physical laws applied to this period of creation.

Steven Weinberg's classic *The First Three Minutes* describes the expansion of the universe containing protons, neutrons, electrons, photons, and neutrinos in complete thermal equilibrium, starting at a much lower temperature of 10^{11} K at approximately 0.02 sec after the big bang. The sequence of events in Weinberg's cosmological chronology is shown in Table 20.2.

It will be helpful for the reader to review the sequence of events that supposedly took place after the explosion of the singularity, in order to assess the Velan model of creation described later.

In both variations the description of events in the early universe, as shown in Tables 20.1 and 20.2 and Figure 20.1, is similar except that Weinberg considers the possibility of the early appearance of protons and neutrons without the intermediate creation of free quarks u and d, which only later were confined into protons and neutrons, when the temperature dropped below 10^{13} K. However, the majority of cosmologists today believe that 10^{-30} sec after the big bang, when inflation ended, the universe consisted of a hot electron–quark soup. According to this scenario, at that fraction of time after the big bang, the strong force, weak

Table 20.1. The Evolution of the Universe in the Classical Singularity Theory Starting at 10^{-45} sec after the Big Bang

Event	Time	Radius	Temperature	Description of event
0	0	Singularity, 0	Infinite	Origin of particles and radiation unknown
1	10^{-45} sec	10^{-50} cm	10^{32} K	Planck era—all forces unified, energy level 10^{19} GeV, no particles
2	10^{-35} to 10^{-30} sec	10^{-33} cm	10^{28} K	Inflation for 10^{-5} sec, universe expands to 10^{-15} cm, origin of matter—no explanation. Gravity acts, three forces united
		10^{-15} cm (size of a proton)		
3	10^{-30} to 10^{-5} sec	0.8 cm	10^{15} K	Gravity, strong and weak force acting. A hot soup of quarks, electrons, photons, and neutrinos in thermal equilibrium
4	10^{-5} to 1 sec	1 cm	10^{13} K	Quarks combine to make neutrons and protons, free electrons. Radiation is still trapped
5	2 sec	6×10^{10} cm	10^{10} K	Neutrinos uncouple from matter and move on with speed of light
6	3 min	1.8×10^{12} cm	10^{9} K	Nucleosynthesis. Neutrons and protons combined into nuclei of hydrogen, helium, and, deuterium
7	700,000 years	6.5×10^{23} cm	3000 K	All protons, neutrons, and electrons in hydrogen atoms (75%), helium (24%), deuterium (1%). Radiation decouples, universe transparent
8	Today, 10–15 billion years	12.47×10^{27} cm	3 K	Fluctuations in density of matter formed protogalaxies, which later evolved into galaxies and stars

The History of Evolution of the Early Universe

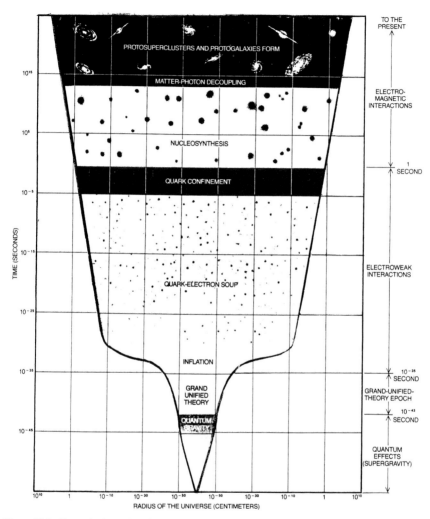

Figure 20.1. Events in the early universe following the classical theory of the big bang and inflation out of a singularity. (Adapted from *Scientific American*)

force, and electromagnetism—all forces of nature except gravity—were unified in a single force. For a short period of 10^{-5} sec, the "universe" (radius of 10^{-33} cm) entered into a special state called false vacuum, in which gravity for just a moment became a repulsive rather than an attractive force allowing the tiny universe to undergo an enormous expansion called inflation. A region that started with a pointlike area, a million, trillion times smaller than an electron, ended up

Table 20.2. The Evolution of the Universe as Described in Steven Weinberg's Book *The First Three Minutes*

Event	Time	Radius	Temperature	Description of event
0	0	Singularity, 0	Infinite	State of infinite temperature and density
Early	0–0.0108 sec		10^{32} K and cooling	Photons, electron–positron pairs, neutrino, and many hadrons all heavily interacting
1	0.0108 sec	Circumference ~4 light-years	10^{11} K	Electrons–positrons, photons, neutrinos. One proton or neutron for 10^9 photons
2	0.12 sec		3×10^{10} K	Electrons–positrons, photons, protons, neutrons, neutrinos (38% neutrons/62% photons)
3	1.21 sec		10^{10} K	Neutrinos decouple. Photons, electrons, positrons (24% neutrons, 76% protons)
4	13.83 sec		3×10^9 K	Electrons–positrons disappear (annihilation), photons. Nucleosynthesis starts
5	3 min 46 sec		10^9 K	Photons, neutrinos (free), electrons–positrons nearly all gone except one electron left per proton. Nucleosynthesis 25% helium, 75% protons
6	34 min 40 sec		2×10^8 K	Photons, neutrinos (free). Nucleosynthesis completed. One electron left for each proton
7	700,000 years		3000 K	Radiation decouples. Nuclei combine with electrons and form 25% helium–75% hydrogen gas
8	Less than 20 billion years	Circumference ~125×10^9 light-years	3 K	Present universe

less than 1 cm (0.8 cm) in size an instant later and became our universe. According to this so-called classical theory recognized by the great majority of the world's leading cosmologists, matter was created virtually out of nothing.

This theory, in addition to boggling the mind, puts under serious scrutiny the credibility as to the judgment of the leading cosmologists, sometimes called "the greatest geniuses in the world."

The theory, which is based on extrapolation of Einstein's equations describing the expanding universe to time zero, was considered by Einstein himself in his later years as lacking credibility. He simply declared that his equations did not apply to the birth and early universe when temperature and energy levels were extremely high.

The theory violates many of the conservation laws of nature such as the conservation of energy, electrical charge, baryon and lepton numbers.

There is also no agreement as to the contents of the singularity. Did it consist of radiating energy alone, compressed to infinite density and temperature, that exploded and transformed part of the energy into particles of matter, or did it contain all of the matter and radiation of the universe in the form of a cosmic black hole, created by the implosion of a previously existing universe? Was there any space around the area of the singularity in which it exploded, or was space created simultaneously during the expansion of the tiny point into a universe?

While writing the last paragraphs of this book and outlining my own complete story of the birth of the universe, the world's leading cosmologists and astrophysicists met in Sweden and Italy for a symposium on the "Birth and Early Evolution of Our Universe." Though skeptical at times, the majority of the scientists reaffirmed their belief in the singularity–big bang creation model, which I consider bizarre, unscientific, and part of metaphysics.

I hope that my theory, the first complete story of creation, developed by an engineer operating outside the professional circles of astrophysicists and cosmologists, will not encounter vehement opposition from the great minds.

20.1. THE YOUNG UNIVERSE AFTER EXPLOSION OF THE FIREBALL IN THE VELAN COSMOLOGICAL MODEL

20.1.1. Just after Explosion

The temperature of the hot electron–quark–gluon plasma, just after explosion, was in the range of 10^{26} K and was rapidly cooling down, inversely proportional to the rate of expansion of the universe, or radius R ($T \sim 1/R$) (Figure 20.2, page 230). In other words, when the universe was a billion times hotter than it is at present (3 K), it was a billion times smaller. For each three quarks and one

THE UNIVERSE JUST AFTER EXPLOSION OF THE FIREBALL

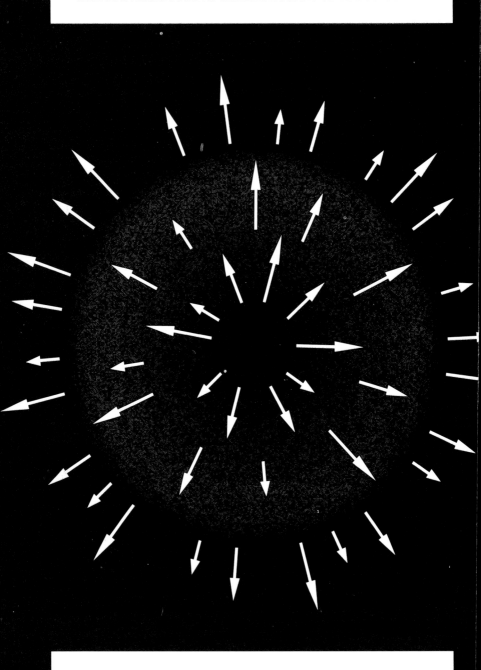

TEMPERATURE JUST AFTER EXPLOSION $10^{26°}$ K.
A DENSE, OPAQUE SOUP OF ELECTRONS, QUARKS,
PHOTONS AND NEUTRINOS IN COMPLETE THERMAL
EQUILIBRIUM. RADIATION DOMINATES.

electron, there were approximately 10^9 photons. Radiation dominated and made the fireball nontransparent, foggy, and opaque. All particles intensively collided with each other and scattered the photons and neutrinos, preventing their escape. The photons moved with the speed of light between collisions with electrons and quarks and scattered. Nevertheless, the radiation redshifted with the expansion of the universe. Due to the predominant quantities of photons and frequent collisions, all particles had the same temperature and the universe was in full thermal equilibrium.

The three forces of nature—the weak and strong nuclear forces and electromagnetism—acted as one unified force.

The free quarks were very close to each other and the temperature was too high for the gluons and the strong nuclear force to contain them within protons and neutrons. Also, at this high temperature and pressure, protons and neutrons would be severely squeezed and quarks would have been released from confinement, a process that takes place in the core of neutron stars and during the collapse of a large star core into a black hole.

The major conservation laws were respected during the creation process and in the expanding, hot universe, in full thermal equilibrium.

20.1.1.1. Conservation of Energy

The total energy of all particles never changed, though collisions transferred energy from one particle to another. The energy of the primordial cosmic radiation field has been partially transformed into particles of matter. A portion of the radiation was trapped and became an important part of the newly created universe. Today, it has cooled down to 3 K and lost its dominating role overtaken by matter when the universe cooled down to 3000 K. Other conservation laws were also conserved.

20.1.1.2. Conservation of Electrical Charge

Today, the universe is electrically neutral. There are as many positively charged protons (+1) as negatively charged electrons (−1). In the early universe after the short annihilation period between electron–positron and quark–antiquark pairs, there was a balance between electrons (−1) and quarks u ($+\frac{2}{3}$) and d($-\frac{1}{3}$) to maintain the electrical neutrality of the young universe.

$$1 \text{ e}^- \text{ to } 2 \text{ u and } 1 \text{ d}$$
$$-1 + (2 \times \tfrac{2}{3}) + (-\tfrac{1}{3}) = 0$$

20.1.1.3. Conservation of Baryon Number

A baryon number of +1 is given to protons and neutrons while leptons and photons have a baryon number of 0. Antiprotons have a baryon number of −1. The significance of the baryon number, which does not create an electrical, magnetic, or similar charge, lies in the requirement to be conserved in interactions of particles.

The quarks u and d, which were contained in the early universe, have a baryon number of $+\frac{1}{3}$. There were three quarks (baryon number +1) in the hot electron–quark soup for each proton and neutron formed later, and so, during the transformation of quarks into protons and neutrons, the baryon number was conserved.

20.1.1.4. Conservation of Lepton Number

At the creation process, there were as many electrons (lepton number +1) as protons (lepton number −1) and neutrinos (lepton number +1) as antineutrinos (lepton number −1), with the result of a 0 lepton number during the birth of the universe. Quarks and hadrons have a 0 lepton number. Later on, after the short annihilation process where only electrons remained, we assume that the distribution between neutrinos and antineutrinos was such that for each electron (+1) there was an electron-antineutrino (−1) and the remaining neutrinos and antineutrinos existed in equal quantity, so the final lepton number count was 0 before and after the annihilation period.

20.1.2. Events in the Fast-Expanding and Cooling Universe

In order to determine the type of interactions and events that took place in the fast-expanding and cooling universe, we must establish the relationship between temperature, density, and energy of the particles. The predominant role in the hot, early universe was played by the energy of radiation, which changes with the fourth power of temperature (T^4) and was considerably greater than the energy contained in the particles of matter, up to 3000 K when matter took over.

The energy of particles of matter relates to the Einstein formula $E = mc^2$ and is expressed in the so-called rest mass: 0.938 GeV for protons, 0.312 GeV for quarks u and d, and 0.511×10^{-3} GeV for electrons.

20.1.2.1. Density and Temperature

The behavior of particles and forces in the hot plasma depended entirely on the prevailing density and temperature, and the temperature depended on the size of the universe ($T \sim 1/R$), directly related to the time elapsed from the explosion, time $t = 0$.

The History of Evolution of the Early Universe

We have established that the energy of radiation E_r is

$$E_r = aT^4 \quad \text{(erg/cm}^3\text{)}$$

where $a = 7.564 \times 10^{-15}$ erg/cm^3 K^4. Therefore, density in accordance with the Einstein equation $E = mc^2$ is

$$d_r = \frac{E_r}{c^2} = \frac{aT^4}{c^2} = 0.84 \times 10^{-35} T^4 \text{ g/cm}^3 \quad (20.1)$$

Equation (20.1) establishes the relationship between density and temperature of radiation.

20.1.2.2. Time and Density

We must now establish the relationship between the time elapsed from the big bang and density, which will give us the basis to interrelate time, temperature, density, and energy of particles. We will be using already known formulas from the Hubble expansion laws and the standard Newton formulas of gravitation.

We take a sphere (Figure 20.3) with radius R containing the mass M. The mass can be determined from its volume and density d_1:

$$M = \frac{4\pi R^3}{3} d_1$$

According to Newton's theory, the potential energy PE of the mass M_1 on the rim of the sphere is:

$$\text{PE} = -\frac{M_1 MG}{R} = -\frac{4\pi R^2 d_1 G}{3}$$

where G is the gravitational constant. The velocity v of M_1 according to Hubble's law is

$$v = HR$$

The kinetic energy KE of the motion of M_1 is

$$\text{KE} = \tfrac{1}{2} M_1 v^2 = \tfrac{1}{2} M_1 H^2 R^2$$

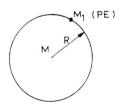

Figure 20.3. A sphere with mass M, radius R, and mass M_1 at the perisphere.

The total energy E_T of M_1 is

$$E_T = -\text{PE} + \text{KE} = M_1 R^2(\tfrac{1}{2}H^2 - \tfrac{4}{3}\pi d_c G) \tag{20.2}$$

If M_1 eventually ceases to expand in a closed universe with slightly more than critical mass, E_T must become 0. Equation (20.2) therefore becomes

$$\tfrac{1}{2}H^2 = \tfrac{4}{3}\pi d_c G \tag{20.3}$$

and we can calculate the critical density d_c

$$d_c = \frac{3H^2}{8\pi G} = 4.5 \times 10^{-30} \text{ g/cm}^3$$

with $H = 15$ km/sec per 10^6 light-years.

From Eq. (20.3) we can establish that

$$H = \left(\frac{8\pi dG}{3}\right)^{1/2} \tag{20.4}$$

As the time after explosion t is reciprocal to the Hubble constant H, we can write

$$t \cong \frac{1}{H} = \left(\frac{3}{8\pi dG}\right)^{1/2} \tag{20.5}$$

We know that the density varies with the radius R or rate of expansion and is inversely proportional to R^3 or $\sim 1/R^3$ for matter and inversely proportional to R^4 or $\sim 1/R^4$ for radiation. We can say that density varies inversely with R

$$d \cong \left(\frac{1}{R}\right)^n$$

where $n = 3$ for matter and $n = 4$ for radiation. Taking Eq. (20.5) we finally arrive at the relationship between expansion time and density:

$$t = \frac{2}{n}\left(\frac{3}{8\pi Gd}\right)^{1/2} \tag{20.6}$$

For the radiation-dominated period of the expanding universe up to a temperature of 3000 K, the final equation is

$$t_r = \tfrac{1}{2}\left(\frac{3}{8\pi Gd}\right)^{1/2} = 0.067\left(\frac{10^8}{d}\right)^{1/2} \quad \text{(sec)} \tag{20.7}$$

For the matter-dominated period of $T < 3000$ K

$$t_m = \tfrac{2}{3}\left(\frac{3}{8\pi Gd}\right)^{1/2} = 0.089\left(\frac{10^8}{d}\right)^{1/2} \quad \text{(sec)} \tag{20.8}$$

20.1.2.3. Energy of Photons and Particles of Matter

In order to determine the energy E_{ph} of a photon in electron volts at a given temperature T, we proceed as follows:

$$E_{ph} = aT^4 = 7.56 \times 10^{-15} \times T^4 \quad (\text{erg/cm}^3)$$

As $1 \text{ eV} = 1.6 \times 10^{-12}$ erg,

$$E_{ph} = \frac{7.56 \times 10^{-15}}{1.6 \times 10^{-12}} \times T^4 = 4.722 \times 10^{-3} \times T^4 \quad (\text{eV}) \quad (20.9)$$

The number of photons N is

$$N = 20.3 \times T^3 \quad (\text{photons/cm}^3)$$

The energy of one photon is

$$E_{1ph} = \frac{E_{ph}}{N} = \frac{4.722 \times 10^{-3} \times T^4}{20.3 \times T^3}$$
$$= 0.232 \times 10^{-3} \times T \quad (\text{eV}) \quad (20.10)$$

The rest mass energy E is

Proton	0.939 GeV
u or d quark	0.313 GeV
Electron	0.511×10^{-3} GeV

We have now established all of the required formulas to calculate time, density, and energy of radiation and particles at a given temperature and can proceed to describe the major events that occurred in the early expanding universe after explosion.

20.1.3. Major Landmark Events in the History of the Young Universe

It took only a millionth of a second for the temperature of the universe to drop from 10^{26} K to 10^{13} K as the expansion was so substantial. The strong nuclear force uncoupled from the unification with the weak force and electromagnetism and soon will confine the now-free quarks and gluons into protons and neutrons. The energy of a single photon is 2.32 GeV.

$T = 10^{13}$ K
1. Strong nuclear force uncouples

Density [Eq. (20.1)]	d	$= 0.84 \times 10^{-35} \times 10^{52} = 0.84 \times 10^{17}$ g/cm³
Elapsed time [Eq. (20.7)]	t	$= 0.067\left(\dfrac{10^8}{d}\right)^{1/2} = 1.4 \times 10^{-6}$ sec
Energy of 1 photon [Eq. (20.10)]	E_{1ph}	$= 0.232 \times 10^{-3} \times T = 0.232 \times 10^{10}$ eV $= 2.32$ GeV
Energy of proton	E_{pr}	$= 0.939$ GeV $\quad E_{1ph} > E_{pr}$

$T = 10^{12}\,\text{K}$

2. Quarks combine to create protons and neutrons

Photons dominate universe

$d = 0.84 \times 10^{-35} \times 10^{48} = 0.84 \times 10^{13}\,\text{g/cm}^3$

$t = 0.067\left(\dfrac{10^8}{d}\right)^{1/2} = 5 \times 10^{-4}\,\text{sec}$

$E_{1ph} = 0.232 \times 10^{-3} \times T = 0.232 \times 10^9\,\text{eV}$
$\qquad\qquad = 0.232\,\text{GeV}$

(There are 10^9 photons to 1 proton or 3 quarks. E_{phT} = energy of 10^9 photons.)

$$E_{phT} = 0.232 \times 10^9\,\text{GeV} \;>\; E_{pr} = 0.939\,\text{GeV}$$

The universe is still in perfect thermal equilibrium filled with radiation and particles of matter in the form of electrons, a quark–gluon plasma, photons, and neutrinos. From the vast spaces of the cosmos, our universe at this stage would look like a nontransparent ball of fog.

The three forces of nature, unified at the high level of energy (10^{15} GeV) and

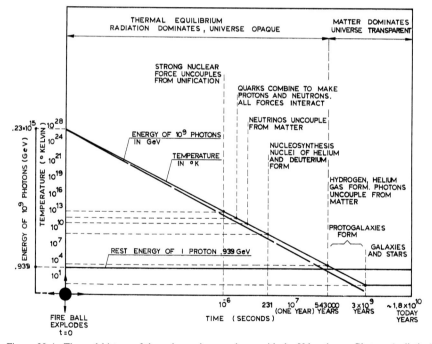

Figure 20.4. Thermal history of the universe in accordance with the Velan theory. Photons (radiation) dominated the universe until the temperature dropped to 40×10^3 K. Rest energy of one proton is 0.939 GeV. The equivalent energy of photons (10^9 photons to 1 proton) was 0.23×10^{15} GeV at $T = 10^{28}$ K and dropped to 0.069 GeV while the rest mass of the proton remained 0.939 GeV. Matter dominated the universe. Radiation uncoupled.

The History of Evolution of the Early Universe

temperature (10^{22} K) that prevailed in the young fireball, came fast into their own lives. The strong nuclear force, increasing in its strength, uncoupled from the unification at a temperature of 10^{13} K, 10^{-6} sec after the big bang and its force was sufficiently high to combine all quarks into protons and neutrons at a temperature of 10^{12} K. The energy level dropped by that time to 100 GeV and the electromagnetic and weak nuclear force uncoupled from the unified electroweak force. The uncoupling of the three forces from unification and its relation to the thermal history of the universe is shown in Figure 20.5.

The freely born quarks u and d in the fireball did not enjoy their asymptotic freedom for very long. They were free when their separation distances were less than 2×10^{-14} cm. Quarks as color singlets with specific charges and gluons responsible for forces acting between them could act as free particles. At high levels of energy and temperature, as shown in Figure 20.5, the strong nuclear force was weaker. Now at temperatures of 10^{12} K, due to the rapid expansion, quarks came within 10^{-13} cm and even if they passed each other at close to the speed of light they could not escape confinement. Interactions take place as fast as 10^{-23} sec and the entire quark–gluon plasma is being transformed into protons and neutrons. Three quarks in three different color singlets become confined into "white," colorless protons and neutrons.

The strong nuclear force of confinement, effected by gluons, is very

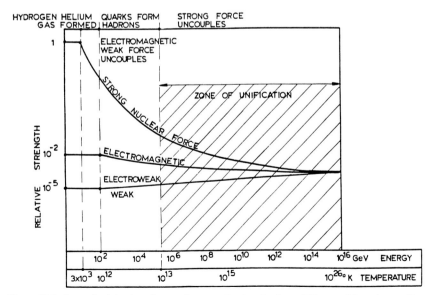

Figure 20.5. Strength of the three forces of nature related to the thermal history of the universe. Strong force uncouples at 10^{13} K, weak and electromagnetic forces uncouple at 10_{12} K.

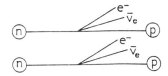

Figure 20.6. Decay of neutrons in beta radiation process.

powerful. The force is approximately 15 tons strong in comparison to 10^{-11} ton for the electric force attracting an electron to a proton in a hydrogen atom. Imagine 15 tons acting on a small pointlike particle such as a quark with a radius of $\frac{1}{3} \times 10^{-13}$ cm. All quarks were confined in protons and neutrons and the content of the universe was transformed suddenly from a quark–gluon–electron–photon–neutrino plasma to a universe of protons, neutrons, electrons, photons, and neutrinos. It is assumed that at this time, much less than 1 sec after the explosion of the fireball, neutrons and protons appeared in equal numbers. However, there was a continuous transmutation of both nuclear particles into each other in weak nuclear force reactions:

$$p + e^- \leftrightarrow n + \nu_e \quad \text{or} \quad n \rightarrow p + e^- + \bar{\nu}_e$$

As free neutrons have a half-life of only 15 min, more and more neutrons decayed into protons. There were basically two different transmutations of neutrons: (1) the classical, so-called beta radioactive decay where two neighboring neutrons produce a proton, electron, and antineutrino as shown in Figure 20.6, and (2) the neutrinoless transformation where the neutrino from the first neutron is absorbed by the second neutron, as shown in Figure 20.7.

The universe still appears as one unified cloud of matter and radiation, nontransparent and completely opaque, though the uniformly appearing homogeneous mass already has some local density fluctuations or seeds of protogalaxies. Its dense fog appearance is caused by scattering of photons by free electrons.

When a free electron is hit by an impinging photon, it is accelerated by the

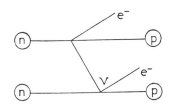

Figure 10.7. Neutrinoless decay of neutrons.

pulse of electromagnetic energy of the photon and gains momentum, as shown in Figure 20.8. The photon loses some energy and momentum, resulting in a change of direction in propagation of electromagnetic energy or scattering of radiation. This process keeps the entire universe in thermal equilibrium. Radiation, which has more energy than particles of matter, dominates and cannot escape.

$\underline{T = 10^{11}\,K}$
3. Neutrinos uncouple

$d = 0.84 \times 10^{-35} \times 10^{44} = 0.84 \times 10^9 \text{ g/cm}^3$

$t = 0.067\left(\dfrac{10^8}{d}\right)^{1/2} = 4.7 \times 10^{-2} \text{ sec}$

$E_{1ph} = 0.232 \times 10^{-3} \times T = 0.232 \times 10^8 \text{ eV}$

$E_{phT} = 0.232 \times 10^8 \times 10^9 = 0.232 \times 10^{17} \text{ eV}$

$\phantom{E_{phT} = 0.232 \times 10^8 \times 10^9} = 0.232 \times 10^8 \text{ GeV}$

$E_{phT} = 0.232 \times 10^8 \text{ GeV} > E_{pr} = 0.939 \text{ GeV}$

At 10^{11} K, 4.7×10^{-2} sec after explosion when the radiation density dropped to 0.84×10^9 g/cm^3, neutrinos and antineutrinos, which were in full thermal equilibrium with matter and radiation until this time, uncouple and move out at the speed of light. These elementary particles with energy and spin, but zero or possibly a small rest mass of 15–30 eV, interact very little with matter. There were and still are 10^9 neutrinos for every nuclear particle in the universe. They have lost considerable energy during the expansion of the universe and may now have an energy of only 0.001 eV or an equivalent temperature of less than 1.5 K. It is for this reason that they are very difficult to detect.

Though the neutrinos ceased to play an active role in particle interactions, their energy continued to contribute to the overall gravitational field of the universe.

$\underline{T = 10^9\,K}$
4. Nucleosynthesis creation of helium nuclei

Nucleosynthesis. Protons and neutrons collide and create nuclei of deuterium and helium. Free protons remain.

$d = 0.84 \times 10^{-35} \times 10^{36} = 8.4 \text{ g/cm}^3$

$t = 0.067\left(\dfrac{10^8}{d}\right)^{1/2} = 231 \text{ sec}$

$E_{1ph} = 0.232 \times 10^{-3} \times T = 0.232 \times 10^6 \text{ eV}$

$E_{phT} = 0.232 \times 10^6 \times 10^9 = 0.232 \times 10^{15} \text{ eV}$

$\phantom{E_{phT} = 0.232 \times 10^6 \times 10^9} = 0.232 \times 10^6 \text{ GeV}$

$E_{phT} = 0.232 \times 10^6 \text{ GeV} > E_{pr} = 0.939 \text{ GeV}$

Figure 20.8. Scattering of photons and electrons.

At 10^9 K, dramatic events take place in the evolution of the universe. This period is called nucleosynthesis and starts to take shape at 231 sec after explosion when free protons and neutrons become bound into atomic nuclei of helium. At the end of this period, lasting approximately 30 min, almost all neutrons ended up in helium nuclei and the final makeup of the massive clouds of matter and radiating energy consisted of 75% protons or nuclei of hydrogen atoms and 25% nuclei of helium atoms, containing two protons and two neutrons. The process of nucleosynthesis, which resembles the thermonuclear reactions taking place in the cores of stars, complies with two basic rules of nature that apply to the formation of atomic nuclei. An atomic nucleus can capture neutrons only one at a time and there cannot be a stable atom with atomic mass 5 or 8. In other words, under normal circumstances, a helium nucleus, which is one of the most stable, cannot capture another proton, neutron, or helium nucleus and form another stable element.

By the time the atomic helium nucleus was created, the density and temperature of the plasma were not sufficiently high for an alternative process where two helium nuclei capture another helium nucleus and combine into carbon ($2\,^4\text{He} + {}^4\text{He} \rightarrow \text{C}$) and later into oxygen ($\text{C} + {}^4\text{He} \rightarrow \text{O}$).

Let us review in more detail the thermonuclear reactions that took place during the nucleosynthesis.

At temperatures of 10^9 K, the collisions between particles, this time between protons, neutrons, and electrons as well as scattering with photons, continued to be fierce. When protons and neutrons came within a distance of 10^{-13} cm, they were subjected to the enormously strong nuclear force. The same force responsible for holding quarks firmly in protons and neutrons by the massless gluons, extended its sphere of influence to hold a proton and neutron in a nucleus of heavy hydrogen called deuterium. The nucleus of deuterium does not have a very strong bond between the proton and neutron. The temperature of the plasma had to be exactly right to avoid blasting apart the deuterium nucleus.

At a temperature slightly lower than 10^9 K, the bond between the proton and neutron in deuterium nuclei became strong and the creation of heavier nuclei became possible. The nuclei of deuterium readily capture neutrons. It is for this reason that heavy water or DHO (water enriched in deuterium) is used in nuclear reactors to absorb neutrons.

The deuterium nucleus colliding with other particles can capture a proton and create a nucleus of the light isotope of helium-3 (^{3}He), which consists of two protons and one neutron, or it can capture a neutron and create a nucleus of the heaviest isotope of hydrogen called tritium (^{3}H), which consists of one proton and two neutrons.

Finally, ^{3}He collides with a neutron and creates the stable helium-4 (^{4}He), or a nucleus of ^{3}H collides with a proton and also creates ^{4}He. Alternatively, a nucleus of ^{3}He collides with another nucleus of ^{3}H, creating ^{4}He and two protons.

As well, a nucleus of tritium can collide with another nucleus of tritium and create ^{4}He and two neutrons.

The nucleosynthesis process is shown schematically in Figure 20.9.

Since helium requires equal numbers of protons and neutrons, the formation of helium stopped when all neutrons were used up. The universe consisted of approximately 25% helium nuclei, 75% free protons, with traces of deuterium nuclei, electrons, photons, and neutrinos. There is obviously one electron present for each free or bound proton. The temperature was still much too high for nuclei to capture electrons and form atoms. Fast-moving photons would knock out the

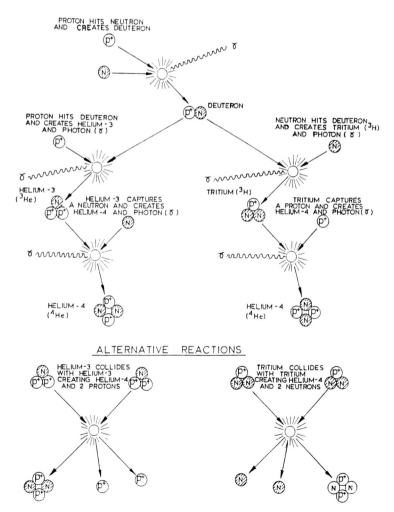

Figure 20.9. The reactions of nucleosynthesis

electrons. Nuclei of heavier elements were not created during the nucleosynthesis as the universe was steadily expanding and cooling down. It remained a hot soup of matter and radiation, still in full thermal equilibrium.

The universe continued to expand and nothing eventful took place for nearly 543,000 years, when the temperature dropped to 3000 K.

$T = 3 \times 10^3$ K

<u>5.</u> Decoupling of radiation, creation of hydrogen and helium gas

$d = 0.84 \times 10^{-35} \times 81 \times 10^{12} = 6.8 \times 10^{-22}$ g/cm^3

$t = 0.067 \left(\dfrac{10^8}{d}\right)^{1/2} = 0.0163 \times 10^{15}$ sec $= 543,000$ years

$E_{1ph} = 0.232 \times 10^{-3} \times T = 0.232 \times 10^{-3} \times 3 \times 10^3$ eV $= 0.69$ eV

$E_{phT} = 0.69 \times 10^9 = 690 \times 10^6$ eV

$E_{phT} = 690 \times 10^6$ eV $< E_{pr} = 939 \times 10^6$ eV

The energy of 10^9 photons corresponding to the energy of 1 proton was now for the first time smaller than the energy of matter in the universe.

At this stage, a significant event took place in the evolution of the universe. The temperature dropped to a level that made the particle–photon scattering process lose its effect and finally electrons and nuclei could form stable atoms. Protons captured electrons and bound them through the electrodynamic force, in compliance with quantum electrodynamics, into atoms of hydrogen and nuclei of helium-captured electrons, and created stable atoms of helium. Suddenly the entire universe was transformed into a ball of hydrogen (75%) and helium gas (25%) with traces of deuterium.

Radiation, which became, for the first time, less energetic than matter, thinning out with the fourth power of expansion and not being scattered by free electrons, suddenly decoupled from matter and escaped with the speed of light. The universe became transparent to radiation, and matter, though still closely packed in one mass, became dominant. Local mass concentrations became more pronounced and the universe suddenly became a red supergiant with a brilliant red light in every part of the sky.

Every point of the universe at this time glowed with the brilliance of the sun. The decoupling or changeover from the domination of radiation to the domination of matter radically changed the behavior of matter. The small density fluctuations in the early stages now became gravitationally very important. The enormous masses of gravitating hydrogen and helium gas started to break up into individual giant gas clouds, slowly drifting apart as the universe continued its fast expansion.

A contributing factor to the breakup of the single mass into individual gas clouds was the sudden drop in pressure when radiation decoupled and moved away with the speed of light. With this development, the radiation temperature continued to drop as the universe expanded and it is presently at 3 K. The universe expanded at this time 1000-fold for the radiation temperature to fall from 3000 K to 3 K. Obviously, the wavelength of the radiation also expanded by a factor of 1000. Matter, after decoupling, cooled much faster than radiation, as the random motion of the atoms of hydrogen and helium could no longer keep up with the expansion of the universe and the larger and larger distances between the individual atoms.

In accordance with the theory of relativity, the photons moving at the speed of light lost the acquired energy slower than did particles of matter, now atoms of hydrogen and helium gas, which moved at a lower speed. Matter was rapidly losing its heat energy and had achieved the 3 K temperature, approximately 1 billion years after expansion and should, presently, theoretically have a temperature of less than 1 K. It is, however, possible that due to outside sources of heat and energy during the formation of galaxies and stars, matter heated up slightly.

Using contemporary theoretical physics, interpreting observational astronomical data, and contributing with my own vision expressed in this new theory of creation, we were able to go back in time and witness the actual birth of our universe in a cosmos containing other similar or perhaps different types of universes and returned to the time when simplicity and symmetry still prevailed in the universe—no structures, only hydrogen and helium gas, photons, and neutrinos. The evolution of the universe from birth to the present is shown in Figure 20.10. (See foldout following page 244.)

At this time I would like to pay tribute to the great spirits, talents, and creativity of so many physicists, astronomers, astrophysicists, and cosmologists who delved into the secrets of the universe, discovered the intricate laws and whose knowledge I have used, adapted, and interpreted in this book. Full acknowledgment and listing of their names is not possible.

In the description of the history of our universe, I went farther than anyone has ever ventured before but it is natural that we would like to go back even farther than the actual birth of our universe and the environment that enabled the process of creation to take place.

The only question that remains is who created the primordial cosmic radiation flowing through the cosmos—a hypothesis introduced in my theory—and going even further we could ask what there was before the cosmos and before the existence of the primordial radiation. In science we leap forward, step by step, and I am sure there will be many attempts in the future to expand my theory with far more explanatory power and maybe even bring about the full understanding of the universe and its purpose.

At this time, I would like to complete this story of creation and enter into the phase of enormous complexity and beauty that started when simplicity in the form of elementary particles turned finally into two basic gases of hydrogen and helium and then, due to the great diversity of interactions, became a base for the development of complex structures and objects that characterize the present universe.

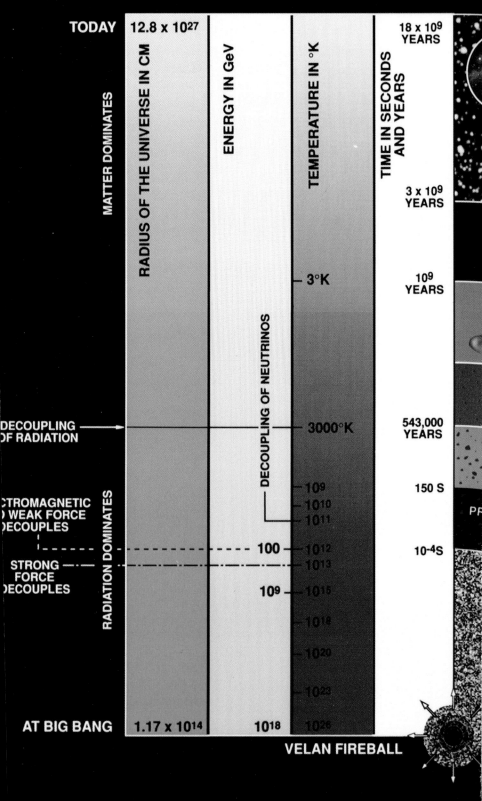

F THE CREATION OF OUR
LTI-UNIVERSE COSMOS

KAREL VELAN, MM, ENG.
1990

S OF EXPANSION FROM THE CREATION
N OF THE PRIMORDIAL FIREBALL OF
-PARTICLES AND RADIATION

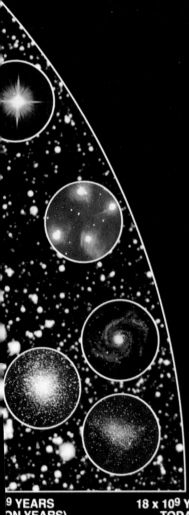

9 YEARS　　　　　18×10^9 YEARS
ON YEARS)　　　　　　TODAY
　　　　　　　(18 BILLION YEARS)

21

Galaxy Formation

The glory of creation continued in its preset manner after the decoupling of radiation when the universe in the form of the expanding fireball cooled down from the unimaginably high level of 10^{26} K to a cool 3000 K. At this stage the universe was still a very simple, expanding ball of gas, 70% hydrogen, 29% helium, and 1% deuterium gas confined to a relatively small area with a radius of 12.47×10^5 light-years and a density of 6.8×10^{-22} g/cm^3. Atoms heavier than helium did not exist. From here on, however, the universe grew continually more complex into protogalaxies, evolving later into galaxies. The principal evolution of matter took place much later in the cores of stars which condensed due to gravitational forces from galactic gas.

In our analysis so far, we have been moving closer to the innermost secret of the universe, its creation. We have witnessed first the transformation of the primordial cosmic radiation into elementary particles of matter at enormously high levels of energy, density, and temperature, the collapse of the mass of matter and radiation into a fireball, its explosion caused by the overwhelming thermal forces, and rapid expansion.

The expansion of the fireball resulted in considerably reduced levels of energy and temperature, the formation of nucleons from quarks, finally evolving into atoms of hydrogen and helium gas. The decoupling of radiation at an approximate temperature of 3000 K and resulting sudden and substantial drop of pressure in the fireball caused the implosion and even collapse of large areas, creating shock waves which in turn compressed the gas.

The Velan fireball during the implosion, as described in Chapter 18, was a homogeneous hot plasma of electrically charged particles such as electrons and quarks u and d and neutral particles such as photons and neutrinos. After the explosion and expansion, when the temperature decreased to 10^{12} K and the energy to 100 GeV, the electromagnetic forces decoupled causing electromagnetic currents which moved around the electrically charged electrons and quarks, causing local irregularities in densities of 2–5% of the otherwise homogeneous plasma.

During the decoupling of radiation, the pressure dropped by a factor of 10^9, as until now the fireball contained 10^9 photons for each baryon of matter. The original irregularities with densities of 2–5% above the average densities of the surrounding areas caused the local gravitational field to increase, resulting in amplifications of densities. Protogalaxies started to form in the time frame of 1 million to 1 billion years after the explosion.

Because of the universal gravitation and the decoupling of the overwhelmingly large radiation, matter could no longer remain uniformly distributed over the large volume of the cooled-down fireball. The enormous masses of gravitating hydrogen and helium gas started to break up into individual giant gas clouds, slowly drifting apart as the universe continued its fast expansion. These clouds contained 10^7 to 10^{15} solar masses.

21.1. FORMATION OF PROTOGALAXIES

The decoupling of radiation took place around 543,000 years after the big bang. The universe continued to cool over the next billion years until the temperature reached only a few dozen degrees Kelvin. The individual gas clouds which formed earlier and accumulated sufficient mass by gravitational attraction started to collapse (Figure 21.1).

The speed of collapse of the clouds became highly supersonic. This created turbulence, which would rapidly increase the growth of irregularities and cause

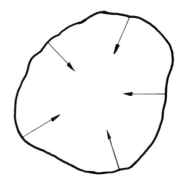

Figure 21.1. A gas cloud forms and collapses under gravitational forces.

Galaxy Formation

fragmentation. The process of fragmentation of the gas cloud led to formation of galaxy-sized fragments or protogalaxies.

The basis for survival of such a fragment was its ability to radiate fast enough the energy in the atoms of gas generated by interatomic collisions. When atoms collide, electrons acquire energy, get excited, and move to higher orbits. However, they instantaneously return to the normal, lowest orbit and radiate away a quantum of radiation in the form of photons. The kinetic energy of the colliding atoms is, in this way, transformed into radiation.

The created radiation can escape from the gas cloud. The denser the cloud, the more collisions, more radiation can escape, and more intensive is the cooling effect of the fragment or protogalaxy. The faster the cloud cools, the more dominant is the gravitational force against the thermal pressure. The gravitational force increases with the size of the collapsing cloud into a protogalaxy. However, the thermal pressure does not depend on size but on temperature and density. Therefore, for each pressure and density there is a minimum mass that will be suitable for gravitational collapse. The fragments that survive and become protogalaxies and later develop into galaxies, must have a so-called minimum Jeans mass where the forces of gravity and thermal pressure are in balance. The Jeans mass is the minimum mass at a given density at which a cloud can condense due to gravitational attraction.

The limiting density is 10^{-22} g/cm^3 and maximum temperature 3×10^3 K. Clouds too hot will start to collapse but will not be able to cool sufficiently fast and the thermal forces will resist the gravitational collapse. This was the condition just before recombination when the universe became dominated by matter.

The minimum (Jeans) mass (M_J) of a cloud that can form by gravitational attraction can be calculated from the assumption that the gravitational potential of mass M must be larger or at least equal to the internal thermal energy of the cloud of mass. The gravitational energy E_G is

$$E_G \approx -\frac{GM^2}{R}$$

where G is the gravitational constant, M the mass of the cloud, and R its radius (Figure 21.2). The internal energy E_G is proportional to the pressure p per unit volume (cm^3). The total energy is proportional to the pressure times volume ($\frac{4}{3}\pi R^3$) or

$$E_{IE} \sim pR^3$$

Therefore, the gravitational contraction of a cloud can take place when

$$\frac{GM^2}{R} > pR^3$$

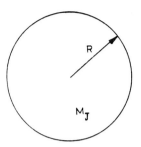

Figure 21.2. Jeans mass M_J, the minimum mass for which gravitational attraction can overcome internal pressure and create a gravitationally bound system such as a galaxy.

or

$$GM^2 > pR^3 \times R$$

$$M = \frac{4\pi}{3} dR^3$$

where d is the density

$$R^3 = \frac{3M}{4\pi d} \sim \frac{M}{d} \quad \text{or} \quad R = \left(\frac{M}{d}\right)^{1/3}$$

$$GM^2 \geq p \frac{M}{d} \times \left(\frac{M}{d}\right)^{1/3}$$

$$M > p \frac{1}{Gd} \times \left(\frac{M}{d}\right)^{1/3}$$

$$M^3 > p^3 \frac{1}{G^3 d^3} \times \frac{M}{d}$$

$$M^2 > p^3 \frac{1}{G^3 d^4}$$

$$M = \left(\frac{p^3}{G^3 d^4}\right)^{1/2}$$

$$\text{Jeans Mass } M_J = \frac{p^{3/2}}{G^{3/2} d^2} \qquad (21.1)$$

At a temperature of about 3000 K, the density was 9.9×10^{-22} g/cm^3. The pressure

$$p = \tfrac{1}{3} d \times c^2 = \tfrac{1}{3} 9.9 \times 10^{-22} \times (3 \times 10^{10})^2$$
$$= 0.3 \text{ g/cm}^3$$

$$M_J = \left(\frac{0.3}{6.67 \times 10^{-8}}\right)^{3/2} \times \left(\frac{1}{9.9 \times 10^{-22}}\right)^2 = 9.7 \times 10^{51} \text{ g/cm}^3$$
$$= 5 \times 10^{18} \, M_\odot \quad (M_\odot = 1.99 \times 10^{33} \text{ g})$$

After the recombination when pressure dropped due to the radiation escape by 10^9, the Jeans mass consequently dropped to

$$M_J \simeq (10^{-9})^{3/2} \times 5 \times 10^{18} \, M_\odot = 1.6 \times 10^5 \, M_\odot$$

Intensive cloud formation started at an estimated density of $d = 10^{-25}$ g/cm^3. At this density the pressure p was

$$p \simeq \tfrac{1}{3}c^2 d = \tfrac{1}{3}(9 \times 10^{20}) \times 10^{-25} = 3 \times 10^{-5} \text{ g/cm}^3 \text{ sec}^2$$

The Jeans mass calculated from Eq. (21.1) is

$$M \simeq \left(\frac{3 \times 10^{-5}}{6.67 \times 10^{-8}}\right)^{3/2} \times \left(\frac{1}{10^{-25}}\right)^2$$
$$\simeq 10^{44} \text{ g} \quad (10^{11} \, M_\odot)$$

This is the average estimated mass of galaxies.

Concerning protogalaxy formation, if the gas cloud is too cool, it will become neutral and little cooling can occur.

Also, smaller fragments may collide with other fragments and be destroyed.

The gas cloud fragments continued to collapse due to gravity, started to spin around their center, and moved at a slower speed than the universe. The internal collapse of the protogalaxies combined with the fast spin caused turbulence and further breaking up. The fragmentation into subfragments stopped when starlike masses were created which collapsed into protostars and the entire cloud or fragment at this stage became a protogalaxy (Figure 21.3).

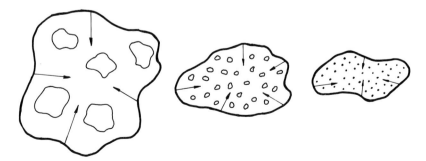

Figure 21.3. The fragmentation of the gas cloud into protogalaxaies and protostars

The rotation of the collapsing clouds was originally influenced by the gravitational pull or torque of neighboring clouds. The rotation can also be attributed to tidal interactions soon after formation and the rotation induced by the fireball to photon "whirls" later transformed into matter. The photon whirls are probably also responsible for seeding small magnetic fields which later amplified with the rapid rotation of the collapsing protogalaxies when rotational speed increased, following the law of conservation of angular momentum.

21.2. FORMATION OF GALAXIES

Once the rotational speed balanced off the gravitational attraction, disklike rotating spiral galaxies were created. The gas was supported by centrifugal forces in the plane of rotation. The top and bottom of the disk flattened because of lack of support and the rotating gas disk fragmented into protostars.

Other protogalaxies that were not exposed sufficiently enough to tidal torques caused by neighboring protogalaxies, rotated at a slower pace and collapsed into oval-shaped elliptical galaxies (Figure 21.4). The further collapse of such galaxies stopped when the orbiting particle clouds rotating around the center mass achieved equilibrium against the total gravitational forces. The formation of galaxies commenced.

Galaxies appeared in the early stages as cold gigantic Frisbees of non-transparent clouds, the largest aggregates of matter known. At this stage, the galaxies contained only hydrogen and helium gas and rotated around their own gravitational centers. At the same time, they were speeding away at up to $0.9c$ or 270,000 km/sec. An estimated 2.5×10^{12} galaxies were formed with an average mass of 10^{11} solar masses, which can be established mathematically as follows. The mass of the universe

$$M_u = 5.68 \times 10^{56} \text{ g}$$
$$1 \text{ solar mass} = 1.989 \times 10^{33} \text{ g}$$

Assuming that 90% of all matter in the universe collapsed into galaxies and the remaining mass is contained in the intergalactic space:

$$90\% \text{ of } M_u = 5.11 \times 10^{56}$$

Assuming the average mass of galaxies is 10^{11} $M_\odot$ or $(1.989 \times 10^{33} \times 10^{11})$ 1.989×10^{44} g, the total number of galaxies in the universe would be

$$\frac{5.11 \times 10^{56} \text{ g}}{1.989 \times 10^{44} \text{ g}} = 2.57 \times 10^{12}$$

Galaxy Formation

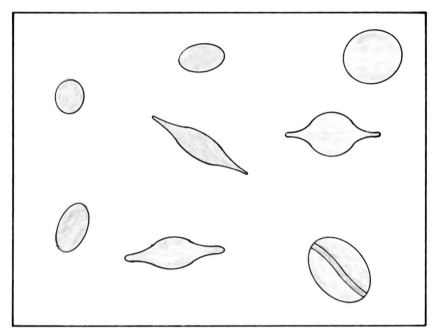

Figure 21.4. Protogalaxies and galaxies as concentrated rotating gas clouds, before creation of any stars: elliptical and spiral structures.

From the possible 2.5 trillion galaxies that may have been created, only 10 billion are presently accounted for in the now-visible universe.

The Velan model of cosmology with its finite and sizable primordial fireball can account for average-scale structures. Also, the modified big bang explosion and prevailing density fluctuations permit large-scale structures such as continuous sheets of galaxies that stretch over enormous areas, giant congregations of galaxies, clusters, quasars, and even clusters of quasars. These large structures, which my theory of creation allows, were discovered recently and are called "great attractors" or "great walls" due to their gravitational influence on other galaxies.

21.2.1. Types of Galaxies

The estimated 2.5 trillion galaxies are now scattered through the universe. Typical galaxies have masses 10^8 to 10^{13} times larger than the solar mass, measure

100,000 light-years across, and contain 100–150 billion stars. Some galaxies, however, contain up to a trillion stars and even more.

21.2.1.1. Classification of Galaxies

In the early 1930s, Edwin P. Hubble organized galaxies in accordance with their optical shape. It took possibly 2–3 billion years after formation of the gas clouds and later protogalaxies before galaxies formed and the stars started to shine and look as we see them today. Before that time, they were invisible clouds of cool hydrogen and helium gas under formation and effects of thermodynamic but mainly gravitational forces. Hubble distinguished between regular and irregular galaxies and created three basic groups—elliptical, normal spiral, and barred spiral galaxies, shown in Figures 21.5–21.8.

Elliptical galaxies (Figures 21.5, 21.6, 21.8) contain little interstellar gas and dust. Most of the original hydrogen and helium gas has now been transformed into mainly old red stars. Their outward shape ranges from nearly spherical systems denoted EO to highly flattened ellipsoids denoted E7. Elliptical galaxies account for approximately 56% of galaxies.

Normal spiral galaxies (Figure 21.7) have bright central regions from which spiral arms extend. S_a are spirals where the central part is very extensive and the arms are tightly wound around it. S_b have a smaller central region and the spiral arms are less closely wound around it, and the S_c have still a smaller nucleus and even looser arms.

Barred spiral galaxies (Figure 21.7) have the two spiral arms starting from the extremities of the central bar of stars. They are divided like normal spiral galaxies into three categories, SB_a, SB_b, and SB_c, depending on the size of the central region in relation to the bars. Spiral galaxies account for 34% of all galaxies.

Irregular galaxies (Figure 21.9) show no symmetry of construction. They account for approximately 10% of all galaxies and owe their geometry to some catastrophic processes in their cores. In many, gigantic black holes are located in their centers.

The irregular galaxy NGC 3034 or M82 (Figure 21.10) shows an enormously large filament of gas erupting out of the galaxy as a result of a gigantic explosion that took place 2 million years ago at the center. The galaxy is a strong source of X rays and most probably a large black hole is located in the nucleus.

In the same category belong the so-called *Seyfert galaxies* named after the discoverer Carl Seyfert. As their cores hold large black holes and have extensive luminosities, they represent a sort of bridge between galaxies and quasars, described earlier in this book. A typical example of this type of irregular galaxy is the exploding galaxy NGC 1275, where filaments of gas are erupting from the center (Figure 21.11).

21.2.1.2. The Real Structure of Spiral Galaxies

Detailed analysis of the movements of stars orbiting spiral galaxies indicates that the optical view alone is insufficient to determine the real structure of the spiral galaxies and their total mass. What we see are only the most luminous objects in the galaxy, the youngest and brightest stars and large clouds of gas illuminated by the blue stars. Contrary to initial views, the spiral arms are not made only out of stars which are distributed quite evenly throughout the galaxy. What we do not see are millions of fainter and dead stars, dark cool clouds of gas, massive black holes at the nucleus of galaxies, comets, planets, neutron stars, molecular hydrogen, remnants of the first generation of stars formed when protogalaxies were initially collapsing, and faint globular star clusters. This enormous mass has a profound influence on the overall gravitational interactions.

What happens here is that density waves ripple through a galaxy and compress interstellar gas along huge spirals creating the so-called emission nebulas or piled-up compressed gas illuminated by young blue stars behind it. Nonilluminated piled-up gas shows up as dark nebulas, usually between glowing emission nebulas. Also, many redder, fainter stars and dead stars are not visible (Figure 12.12).

In actual terms, only 15–25% of galactic material is visible. Spiral arms are embedded in a flattened disk surrounding the central core of the galaxy, in addition to a halo of faint globular clusters, each with 200,000–500,000 stars and a large invisible corona of nonluminous cold material. All of this material keeps the galaxy together and explains that stars, no matter how distant they are from the center of the galaxy, orbit at nearly the same linear velocity, contrary to normal analysis of the influence of the gravitational field.

The gravitational attraction of the material contained in the invisible halo extending far beyond the visible disk obviously pulls on the stars, causing even the stars most distant from the center to orbit at high velocity.

21.2.1.3. Clusters

While still evolving 2 billion years after formation, many galaxies attracted more mass and other galaxies and formed clusters of galaxies and, later on, superclusters. Regular clusters (a typical one is shown in Figure 21.13) contain mostly elliptical galaxies and have a spherical shape.

Irregular clusters have indefinite shapes and contain all types of galaxies. A

Figure 21.5. Elliptical galaxies ranging from (a) spherical geometry to (b–d) flattened systems. Red stars dominate elliptical galaxies with very little interstellar matter. Elliptical galaxies account for approximately 56% of all galaxies. (National Optical Astronomy Observatories)

Galaxy Formation

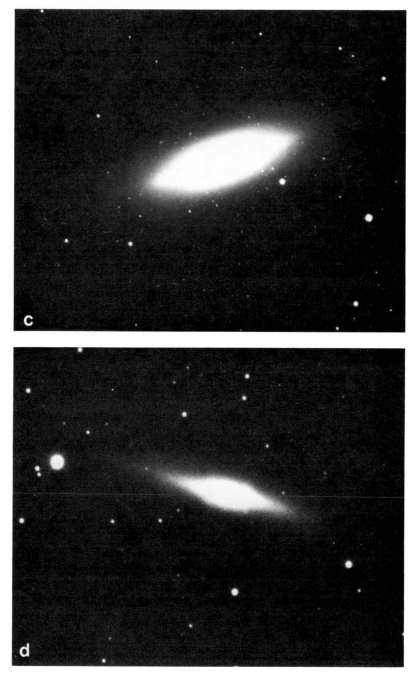

Figure 21.5. (*Continued*)

Figure 21.6. Long exposure of the brightest elliptical galaxy known as M87 NGC 4486 in the Virgo cluster reveals that the galaxy's halo includes over 500 globular star clusters, each containing over 100,000 stars. In the center of the galaxy there seems to be a black hole of 5 million solar masses. (Lick Observatory)

Figure 21.7. Normal spiral galaxies (a, c, e) have a pronounced central core and arms around it, wound closely on (a) and more loosely on (c). In barred spiral galaxies (b, d, f), the arms start from the two extremities of a central bar. Spiral galaxies account for approximately 34% of all galaxies. What we observe is only the most luminous objects. In effect, galaxies contain large dark clouds, halo of faint stars, dead stars, nonluminous matter, and giant molecular clouds. (Royal Observatory, Edinburgh, National Optical Astronomy Observatories, and Lick Observatory)

Figure 21.7. (*Continued*)

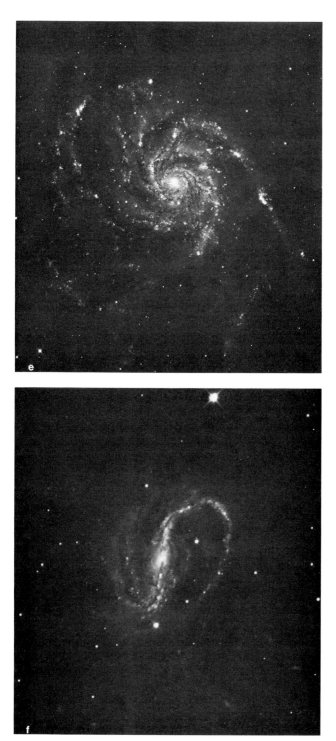

Figure 21.7. (*Continued*)

Figure 21.8. The elliptical galaxy M87 is shown here as part of the Virgo cluster. Photograph taken by the Cesso-Tolato observatory in Chile, operated by the Kitt Peak National Observatory.

Galaxy Formation

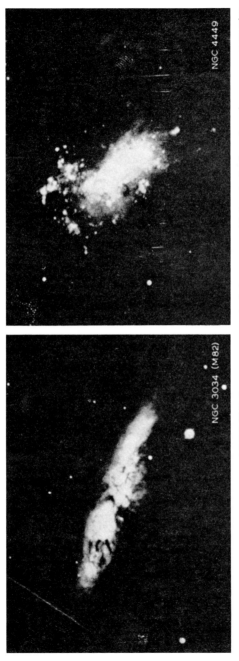

Figure 21.9. The two galaxies NGC 3034 and NGC 4449 show no symmetry in their systems and belong to the irregular galaxy group representing approximately 10% of the total and are the most interesting structure in the universe. (Lick Observatory)

262   Chapter 21

Figure 21.10. The irregular galaxy NGC 3034 or M82 exploded 2 million years ago. (California Institute of Technology)

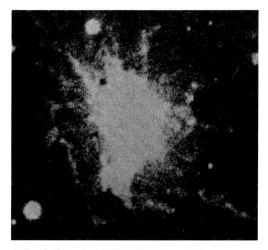

Figure 21.11. The exploding Seifert galaxy NGC 1275. Filaments of gas are erupting from the center of this irregular galaxy with a powerful black hole in the center. (California Institute of Technology)

Galaxy Formation

Figure 21.12. An emission lit up nebula in the constellation of Serpense. A birthplace of stars. (Lick Observatory)

Figure 21.13. Regular cluster of galaxies. Many of the estimated 2.5 trillion galaxies scattered across the universe are grouped in clusters. This regular, usually spherical cluster contains mostly elliptical galaxies. (California Institute of Technology)

typical irregular galaxy is shown in Figure 21.14. It is located in the constellation of Hercules about 350 million light-years away, and contains 100 galaxies, many of them spirals.

We belong to a local group cluster, a small congregation of 24 galaxies, the majority being dwarf elliptical galaxies. The largest galaxy in our own group is Andromeda, a spiral galaxy about $2\frac{1}{4}$ million light-years away and containing about 340 million solar masses. The local group is about 3 million light-years across and is approximately 10 billion years old.

Clusters may contain 20–30 galaxies but some of the largest groupings in superclusters may contain 1000–2000 galaxies and extend over a region as large as 300–400 million light-years in diameter.

Galaxy Formation 265

Figure 21.14. Irregular cluster of galaxies. This cluster, about 350 million light-years away, has an irregular shape about 5 million light-years across. It is located in the constellation of Hercules and contains 100 galaxies, mainly spirals. (California Institute of Technology)

Galaxies in cluster orbit around each other and remain in the same area of space due to mutual gravitational attraction and lack of enough energy to escape from the cluster. There is enough mass to slow down the motion of a cluster against the normal Hubble expansion of the universe by 10–20%.

Giant concentrations of matter or agglomeration of galaxies have been found recently, called "great walls" or "great attractors," extending over 600 million light-years across, which formed when superclusters of galaxies collided or were attracted gravitationally to create this enormous concentration of mass influencing neighboring clusters. The recent discovery of these gigantic galactic superstructures of matter puts in jeopardy the classical theory of the big bang and inflation. Both call for an isotropic and homogeneous distribution of matter in the universe, claiming that inflation, which took place 10^{-35} sec after the big bang and lasted 10^{-5} sec, eliminated all irregularities in the primordial universe, at that time smaller than an atom.

The Velan theory of creation with a finite and large primordial fireball does not exclude density irregularities which may have been responsible for the creation of these enormous "great walls" of matter.

Gravitational behavior of many clusters indicates that they must contain additional nonvisible mass in the form of burned-out galaxies, small dwarf dim galaxies, intergalactic gas and dust, and even newly formed galaxies born within the past 100 million years, when stars are only in their forming stage.

22

The Birth of the First Generation of Stars

Our expanding universe, which we were able to analyze from the time of its creation to the formation of galaxies, was opaque and nontransparent for the first half-million years when radiation decoupled and the temperature dropped to approximately 3000 K. At that time, the universe, still closely packed in one mass though local concentration became more pronounced, became a transparent red supergiant with a brilliant red light in every part of the sky. Every point of the universe at this time glowed with the brilliance of the sun.

After a half-billion years the universe cooled down considerably to a cool 10–50 K and became opaque again. The clouds which developed into protogalaxies and later evolved into galaxies appeared as barely visible gigantic aggregates of matter, all in the form of hydrogen and helium gas.

Three billion years later, when galaxies were still shaping their structures, the ultimate mystery of mass evolution started to unfold, when trillions upon trillions of Population 3 stars condensed from galactic gas (mainly hydrogen) and made galaxies visible for the first time. Magnetic and density waves, rippling through the galactic gas masses, were compressing the cold hydrogen in many areas until lumps of piled-up gas were created. These lumps possessing higher than average gravitational energy attracted more and more gas and grew into a self-perpetuating process called gravitational accretion.

The core of the collapsing mass, unable to support the weight of trillions upon trillions of tons of gas, caused the collapse of the mass until the internal gas pressure was high enough to support the weight of outer layers and a hydrostatic equilibrium was achieved. While some of the outer layers were still falling in, the core was stabilizing and a *protostar* was born. The gravitational energy was being converted into thermal energy and the core of the protostar was progressively heated up to 170,000 K from the cold gas at 3 K.

As convection currents were carrying heat outward and cooler material from the outer layers was sinking down to provide a corrective heat balance, pressure and temperature in the core were dropping, causing the weight of all of the gas to press more than ever. The contracting of the protostar considerably increased the core's temperature. When the temperature reached 4 million K, protons or hydrogen nuclei were compressed so closely and were moving so rapidly, they collided and fused into helium nuclei and helium atomic gas by interaction of the strong nuclear force, a process that took place in the early universe during the nucleosynthesis period.

The thermonuclear burning or fusion of hydrogen into helium was initiated and an enormous quantity of high-energy radiation was being released into space. *This signaled the birth of a star.* As soon as the release of energy resulting from the thermonuclear reaction became large enough to compensate for the energy lost by radiation, the gravitational contraction of the star stopped and thermal equilibrium was established.

The time for a protostar to develop into a hydrogen-burning star, its luminosity and surface temperature depend largely on the size of the mass concentrated in the protostar. It takes approximately 1 billion years for a $\frac{1}{4}$ solar mass to become a hydrogen-burning star, 30 million years for 1 solar mass, 300,000 years for 10 solar masses, and as little as 30,000 years for 50 solar masses. If we take the luminosity of our sun as 1 with its surface temperature of 6000 K, the luminosity of large stars can be as much as 80,000 higher and the surface temperature can reach 35,000 K for a star of 25 solar masses.

In a star like our sun, of which the composite mass is set as 1, the burning of hydrogen into helium will take approximately 10 billion years, during which period 600 million tons of hydrogen is fused into helium every second (Figure 22.1).

In accordance with the so-called virial theorem, in a star in mechanical equi-

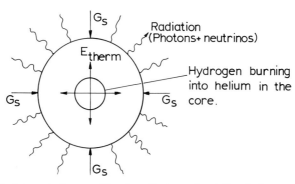

Figure 22.1. Schematic diagram of burning of hydrogen into helium in the core of the sun.

librium, the thermal forces E_{therm} are equal to the gravitational force G_s expressed as follows:

$$2E_{therm} + G_s = 0 \tag{22.1}$$

From this equation we can determine the temperature T of the sun or any other star.

The equation for thermal energy E_{therm} due to motion of particles N (protons + electrons) is for the sun:

$$E_{therm} = 3N\,kT = 3\frac{M_\odot}{M_p}kT \tag{22.2}$$

where k is the Boltzmann constant, N is the number of protons, M_p is the mass of a proton, G is the gravitational constant = 6.673×10^{-8} cm³/g-sec², $M_\odot$ is the mass of the sun, and

$$G_s = G\frac{M_\odot^2}{R_\odot} \tag{22.3}$$

From Eq. (22.2) we obtain the relationship between the temperature and the mass of the star:

$$\frac{M_\odot}{M_p}kT = G\frac{M_\odot^2}{R_\odot}$$

$$3kT = G\frac{M_\odot}{R_\odot}M_p \approx 600 \text{ eV} \tag{22.4}$$

$$T \cong 600 \times 10^4 \text{ K}$$

This is the average temperature. The actual temperature of the core is 20×10^6 K and on the surface of the sun's photosphere, 6000 K. The star's internal temperature depends entirely on the mass of the star, and the gravitational contraction of the star stops as soon as the energy in the nuclear processes (hydrogen being transmuted into helium in the core) becomes large enough to compensate for the energy lost by radiation. Presently, 600 million tons of hydrogen are converted into helium each second at the sun's core.

Originally, at the forming stage the sun was 75% hydrogen and 25% helium. Presently, after 4.5 billion years of thermonuclear reactions, hydrogen has been depleted to 35% and helium has increased to 65%.

22.1. STAR'S LUMINOSITY

The Stefan–Boltzmann law states that the total radiation emitted by a blackbody (we used it in calculating the energy of radiation in the fireball) is proportional to the fourth power of temperature:

$$E = at^4 \quad (\text{erg/cm}^2\text{-sec})$$

where $a = 7.56 \times 10^{-15}$ erg cm^{-3} K^{-4}. The total luminosity of a star, assuming it to be a blackbody (in thermal equilibrium), is found by multiplying the energy emitted in a unit area (cm^2) by the surface area. Therefore, the luminosity L of a star with temperature T and radius R is

$$L = \underset{\text{surface}}{4\pi R^2} \times aT^4 \tag{22.5}$$

$$L = \underset{\text{constant}}{4\pi a} \times R^2 T^4 \tag{22.6}$$

If two stars have the same effective temperature but their radii differ by a factor of 3, then the larger star radiates 4 times as much energy as the smaller star.

We calculate the luminosity of the sun (mass 1) as

$$L_\odot = 3.8 \times 10^{33} \text{ erg/sec} \quad (L = 1)$$

Luminosity is directly related to the second power of the radius of a star and the fourth power of the star's temperature.

22.2. HERTZSPRUNG–RUSSELL DIAGRAM

If we plot the relationship between the surface temperature of a star and its luminosity, we arrive at a Hertzsprung–Russell diagram with a center line called the main sequence equilibrium stage line (see Figure 22.2).

When a star settles down on the main sequence line of the H–R diagram, it starts the longest active phase of its life. During that period, the star is powered by thermonuclear reactions. The heat from these reactions causes the star to shine and maintains a pressure to support the star against the tendency of inward collapse, caused by its own gravitation. In the case of the sun, it is the transmutation of 600 million tons per second of hydrogen into helium in a thermonuclear reaction taking place in its core.

Figures 22.2 shows the main sequence evolution of four protostars. High-mass protostars evolve rapidly. For a star with a mass of 30 solar masses, it takes only 30,000 years to reach the main sequence, while for a star with 1 solar mass, such as our sun, it takes 30 million years (Table 22.1).

During a protostar's formation, due to gravitational implosion and core heating, the surface temperature is constantly changing and consequently the luminosity. Once the thermonuclear reaction is triggered and energy is radiated, the equilibrium is established and the born star arrives at the main sequence line

The Birth of the First Generation of Stars

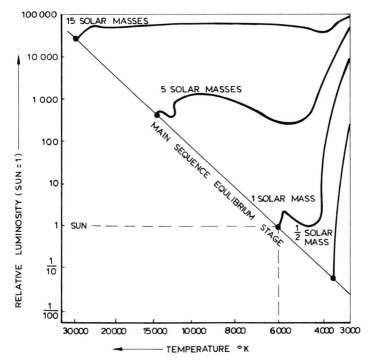

Figure 22.2. Hertzsprung–Russell evolution diagram for four stars of ½ to 15 solar masses until they reach the main sequence equilibrium stage line.

of the H–R diagram. The thermonuclear fusion of hydrogen into helium is shown schematically in Figure 22.3.

All stars are classified in accordance with their luminosity L and surface temperature T. H–R diagrams are shown in Figures 22.2 and 22.4.

Above the main sequence, stars active in thermonuclear reactions are "red

Table 22.1. Time for a Protostar to Develop into a Hydrogen-Burning Star

Mass of protostar (sun = 1)	Time required to reach main sequence
30 solar masses	30,000 years
10 solar masses	300,000 years
1 solar mass	30 million years
½ solar mass	100 million years

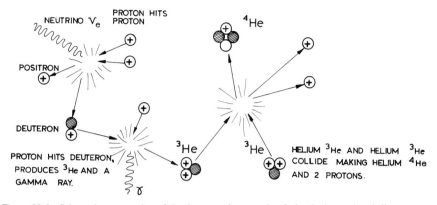

Figure 22.3. Schematic presentation of the thermonuclear reaction fusing hydrogen into helium.

giants," which combine low temperature with high luminosity due to their large dimensions. In the lower corner are white dwarfs with high surface temperature and very low luminosity due to very small size. As an example, the diameter of the Wolf 457 white dwarf is 300 times smaller than the sun. The diameter of the red giant Betelgeuse in the Orion constellation is 450 times larger than the sun. The supergiants have enormous luminosity due to their gigantic surface. As an example, Deneb in the Cygnus constellation has a luminosity 600 times, and a mass 30 times the mass of the sun.

22.3. EVOLUTION OF STARS

Hydrogen and helium gas created in the fireball from protons, neutrons, and electrons remained intact during the formation of protogalaxies which later evolved into galaxies. The formation of heavier elements such as carbon, oxygen, silicon, and iron took place in the center of stars, first created from the galactic hydrogen and helium gas.

The first generation of stars are called Population 3 stars. When the stars started to collapse due to their own gravitation, the temperatures in their cores increased until they reached the so-called thermonuclear igniting level. The lowest temperature is 4 million K and is sufficient to trigger the fusion of hydrogen into helium. The thermonuclear reaction that started with the burning of hydrogen continues in the cores of stars depending on their sizes. Smaller stars of up to 4 solar masses produce only helium, carbon, and oxygen. Stars of 4–6 solar masses fuse, in addition, heavier elements such as nitrogen, neon, sodium, and magnesium. Stars of 6–8 solar masses add to it sulfur and silicon, and stars of

The Birth of the First Generation of Stars

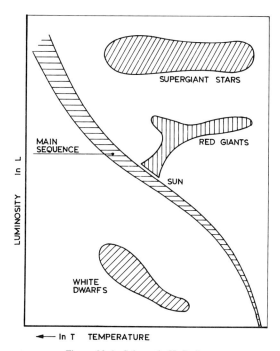

Figure 22.4. Schematic H–R diagram.

more than 8 solar masses complete the cycle with iron. The dying Population 3 stars exploded in supernovas, ejected large quantities of heavy elements into space, and created nebulas or large concentrations of interstellar gas.

These interstellar clouds of gas that were enriched with the heavy elements became the birthplace for the next generation of stars, called Population 2 stars. They contained at least 1% of the heavier elements. Population 2 stars produced, in thermonuclear reactions, heavier elements. Later on in supernova explosions, these elements were spewed into the galactic space factory creating new nebulas. This highly enriched gas, approximately 5 billion years ago, collapsed to form Population 1 stars such as our sun with up to 3% heavy elements.

In all stars the hydrogen-burning time depends on the size and mass of the star. It may last only 3 million years for a massive star of 25 solar masses, or 10 billion years for a small star the size of our sun. Stars smaller than the sun evolve extremely slowly. Hydrogen-burning in a star half the size of the sun would take 200 billion years, exceeding the expected age of the universe. In stars smaller than 0.3 of the sun, no helium core can develop. After 10 billion years, only 1% of the hydrogen is consumed and only isotopes of helium-3 form.

22.4. ACTIVE LIFETIME OF STARS RELATING TO MASS

The active lifetime of a star or the period of thermonuclear activity in its core, which as a rule is about 10% of the total mass, depends entirely on the size of a star.

Large stars, which radiate intensively and have high luminosity, consume the nuclear fuel rapidly and are short-lived. Smaller stars such as the sun radiate at a much slower rate, consume nuclear fuel at a slower rate, and have a long life. During the thermonuclear burning, the star is in hydrostatic equilibrium and the thermal and gravitational energy are in balance. The thermonuclear fuel supply of a star depends entirely on its mass M, in accordance with the Einstein equation $E = Mc^2$.

The mass defect, resulting from transmutation of hydrogen nuclei or proton into helium, is 0.007 and into iron 0.008. The core where thermonuclear reactions take place is approximately 0.10% of the star. Therefore, the total nuclear reserve of a star is

$$E_{nucl} = k \times 0.1 \times Mc^2 \qquad (22.7)$$

where M is the star's mass and k is the coefficient of mass defect or "lost mass," as a result of transmutation of protons (hydrogen nuclei) into other nuclei. From Eq. (22.7) we can estimate the nuclear energy supply of the sun $M_\odot = 2 \times 10^{33}$ g:

$$\begin{aligned} E_{nuc.s} &= 0.007 \times 0.1 \times M_\odot \times (3 \times 10^{10})^2 \\ &= 0.007 \times 0.1 \times 2 \times 10^{33} \times 9 \times 10^{20} \\ &= 0.0126 \times 10^{53} \\ &= 1.26 \times 10^{51} \text{ erg} \end{aligned}$$

For a star of 20 solar masses, the supply of nuclear energy is

$$\begin{aligned} E_{nuc.20} &= 0.008 \times 0.1 \times 40 \times 10^{33} \times (3 \times 10^{10})^2 \\ &= 28.8 \times 10^{51} \text{ erg} \end{aligned}$$

22.5. ACTIVE AGE OF A STAR

From the total nuclear energy reserve E_{nucl} and luminosity L, one can calculate the active age of a star and its presence in years on the H–R main sequence line. Using Eq. (22.6), t_{star} or time for thermonuclear activity of a star is

$$t_{star} = \frac{E_{nucl}}{Lc} \times 3.2 \times 10^{-8} \text{ year} \qquad (22.8)$$

The Birth of the First Generation of Stars

As an example, we can calculate the active life of the sun with luminosity $L = 3.8 \times 10^{33}$ erg/sec and $E_{nuc.s} = 1.26 \times 10^{51}$ erg:

$$t_{mn} = \frac{1.26 \times 10^{51}}{3.8 \times 10^{33}} \times 3.2 \times 10^{-8} \text{ year}$$

$$= 10^{10} \text{ years} = 10 \text{ billion years}$$

As the sun has burned hydrogen for 4.5 billion years, the remaining lifetime is 5.5 billion years.

Luminosity of a star is approximately proportional to the third power of its mass. As already mentioned, depending on the size of the star, the thermonuclear reaction may continue after the hydrogen burning. The star collapses, increasing its core temperature. Once the temperature rises to a level sufficiently high to "ignite" the burning of helium and even heavier elements after, the active life of the star continues and the star remains on the main sequence. The ignition and operating temperatures for burning the various elements are shown in Table 22.2

The rate of liberation of the nuclear energy resulting from the thermonuclear burning in the core of stars is very slow. The energy yield e from the sun, for instance, per gram of mass is only

$$e = 2 \text{ erg/g} \tag{22.9}$$

In spite of the enormous temperature of the plasma of 20 million K in the core of the sun, the surface temperature is only 6000 K.

Obviously, the enormous mass and low conductivity of gas surrounding the burning core provides the thermal insulation, in addition to the gigantic gravitational pressure of the outer layers. It is for this reason that a thermal explosion in the center of the sun will show up on its surface only after several million years.

After the thermonuclear reaction cycle which keeps the star in hydrostatic equilibrium comes to an end, dramatic evolution takes place. Stars of up to 4 solar masses evolve into red giants, discard their outer layers forming planetary

Table 22.2. Thermonuclear Reactions and Star Masses

Type of burning	Minimum star mass (Sun 1)	Ignition temp. (K)	Operating temp. (K)
Hydrogen	$\frac{1}{10}$	4 million	20 million
Helium	$\frac{1}{2}$	100 million	200 million
Carbon	4	600 million	800 million
Neon	6	1.2 billion	1.6 billion
Oxygen	6	1.5 billion	2.1 billion
Silicon	8	2 billion	3.5 billion

Table 22.3 Evolution of Stars

Star mass (sun = 1)	Time in years for protostar before hydrogen burn	Surface temp (K)	Luminosity (sun = 1)	Time (years) for H burning	Core burning H	He	C	O	Si	Final core C	O	Si	Fe	Shell Planetary nebula	Super nova	Core White dwarf	Neutron star	Black hole
25	60,000	35,000	80,0000	3×10^6	✓	✓	✓	✓	✓				✓		✓			✓
15	150,000	30,000	30,000	15×10^6	✓	✓	✓	✓	✓				✓		✓			✓
8	400,000	19,000	5,000	100×10^6	✓	✓	✓	✓	✓						✓		✓	
6	600,000	17,000	700	300×10^6	✓	✓	✓	✓				✓			✓		✓	
4	1 million	15,000	100	400×10^6	✓	✓	✓				✓				✓			
3	4 million	11,000	60	500×10^6	✓	✓				✓	✓			✓		✓		
1½	10 million	7,000	5	3×10^9	✓	✓				✓	✓			✓		✓		
1	30 million	6,000	1	10×10^9	✓	✓				✓	✓			✓		✓		
¾	60 million	5,000	½	15×10^9	✓	✓				✓				✓				
½	100 million	4,000	¼	200×10^9	✓													

The Birth of the First Generation of Stars

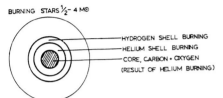

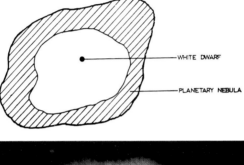

Original Star Mass of Red Giant	Ejected Nebula Mass	Mass of White Dwarf
3.5	2.1	1.4
3.0	1.8	1.2
1.5	0.7	0.8
0.8	0.2	0.6

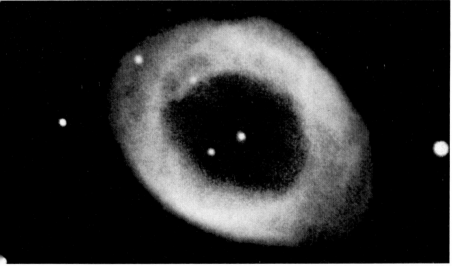

Figure 22.5. Ring Nebula in Lyra (M57) with white dwarf in the center. (California Institute of Technology)

nebulas, and become white dwarfs. Stars of more than 4 solar masses evolve into supergiants, explode, leaving most of their mass in a supernova, while the star's core collapses. When the remaining mass of the core does not exceed 2.5 solar masses, the core becomes a neutron star. When a mass of the naked core exceeds 2.5 solar masses, a black hole is created. A schematic presentation of the evolution of stars is shown in Table 22.3.

22.6. THE FATE OF STARS OF UP TO 4 SOLAR MASSES

The fate of the dying stars depends on their size. Stars of $\frac{1}{2}$ to 4 solar masses continue their active life and remain on the main sequence by burning part of the helium. After the star becomes a red giant, most of the material remains in a planetary nebula and the core of the star shrinks to a white or black dwarf, as shown in Figure 22.5.

A detailed analysis of a typical star in this category, our sun, is presented in the next chapter.

23

The Fate of the Sun and Planetary System

23.1. SUN'S MASS FORMS

Approximately 4.6 billion years ago, a lump possessing higher than average gravitational energy attracted more and more interstellar gas and dust enriched with heavy elements spewed out by supernova explosion, and the process of gravitational collapse was initiated. The core, unable to support the weight of trillions of tons of infalling gas, contracted and the temperature increased from 3 K to approximately 180,000 K, as gravitational and kinetic energy was being converted to heat. At this moment, the internal gas pressure was sufficiently high to support the outer layers and a hydrostatic equilibrium was achieved. This signaled the birth of the protosun (Figure 23.1).

The protosun was a hot and cold blob of gas in continuous movement. As heat from the hot core was rapidly rising, cold masses of gas were falling in toward the center. Convection currents were carrying heat outward and cooler material was sinking down as in a boiling soup. Pressure and temperature of the core were dropping. However, soon after the heat was carried away from the core, the enormous mass of the outward layers compressed the core and rapidly increased its temperature. When the density of the core increased substantially and the temperature reached 4 million K, hydrogen nuclei stripped of electrons were squeezed so tight that they started to fuse. This triggered the thermonuclear fusion process or burning of hydrogen into helium. At this moment the sun was born as a Population 1 star, a ball of hydrogen and helium gas, enriched with 3% heavy elements. The remaining gas consisted of approximately 74% hydrogen and 23% helium.

Once the hydrogen burning was initiated, the star arrived on the H–R

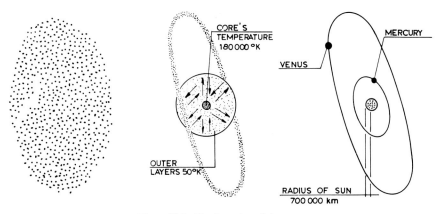

Figure 23.1. The formation of the sun.

diagram as shown in Figure 22.2. This process of formation took about 30 million years. The hydrogen burning, which takes place only in the center of the sun, consumes (per second) 600 million tons of hydrogen which turns into helium.

The final equation for the thermonuclear fusion of hydrogen is

$$4H \rightarrow {}^4He + \text{energy } (E_m)$$

The energy E_m is released mostly in the form of (γ) radiation and electron-neutrinos and can be calculated from:

$$E_m = mc^2$$

where m is the missing mass, accounting for the difference in weight of the four hydrogen nuclei and the resulting nucleus of helium. The radiated energy is called luminosity. The missing mass m is equal to 0.007 of the mass of the four protons. For the sun the total energy is $E_s = 0.007 \times 0.1 \times Mc^2$ as only 10% of the solar mass (0.1) is contained in the burning core. For a detailed calculation, see Chapter 22.

23.2. THE SUN TODAY

After 4.5 billion years, the sun today is a typical medium star. Its mass and luminosity, for comparison with other stars, are considered to be 1, as shown in Figure 22.2.

The major observed parameters of the sun are

Mass $M_\odot$	2×10^{33} g
Radius $R_\odot$	7×10^{10} cm
Mean density	1.4 g/cm^3
"Surface" temperature	6000 K
Radiated energy or luminosity $L_\odot$	3.8×10^{33} erg/sec
Chemical composition today	
Hydrogen	72%
Helium	25%
Heavier elements (carbon, nitrogen, oxygen, neon, silicon, iron, etc.)	3%

Nearly 5 billion years of thermonuclear fusion in the core of the sun depleted some of the hydrogen and increased slightly the amount of helium. The present operating temperature in the core is about 20 million K.

23.3. THE STRUCTURE OF THE SUN

The sun, like most other stars, has a central body called the photosphere, or sphere of light. This is really the sun that we see. It is not the surface of the sun as there are two other layers above the photosphere. The photosphere is the outer layer of the central body from which the visible light is radiated. The temperature of the photosphere is about 6000 K, which rapidly increases toward the center where thermonuclear fusion takes place at 20 million K. The photosphere is opaque, consisting mainly of isotopes of hydrogen, deuterium or heavy hydrogen, a hydrogen nucleus (proton) that has captured a neutron. The structure of the photosphere is granular (Figure 23.2). The granules of hot, boiling gas up to 1000 km in diameter appear continuously for 10–20 min.

The photosphere is a 300-km skin and forms the visible surface or disk of the sun (Figure 23.3). Energy is transported from the core mainly by radiation in the form of photons through the gas in the interior of the sun, which is transparent to radiation, to a radius of about 595,000 km or 85% of the interior. What really happens is that radiation is absorbed and remitted many times on its way out. In the zone of the last 15%, just below the photosphere, the pressure and temperature are such that singly ionized helium ions form (He$^+$) and they can be photoionized. The consequence is that the gas becomes convective and it is opaque.

In this area called the convection zone, the energy is being transported by means of convections. Hot gas columns rise and cool and cool columns sink and

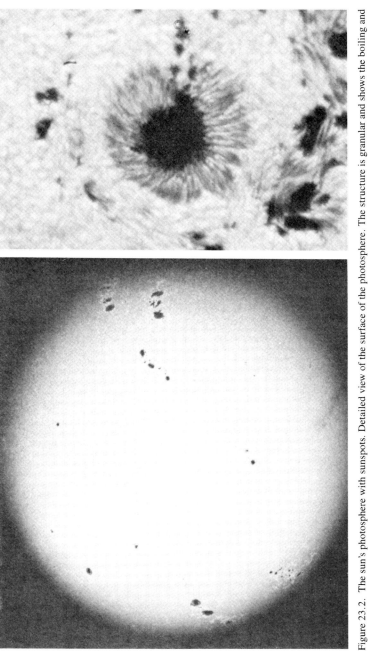

Figure 23.2. The sun's photosphere with sunspots. Detailed view of the surface of the photosphere. The structure is granular and shows the boiling and bubbling of the surface. The granules up to 1000 km in diameter appear for 10–20 min. (California Institute of Technology)

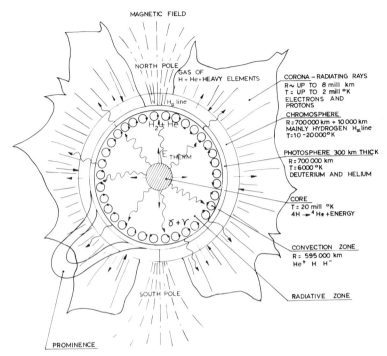

Figure 23.3. Schematic diagram of the sun. Energy is generated in thermonuclear burning core (equivalent to trillions of hydrogen bombs exploding each second) and transported through 85% of the interior by photons (γ). Through the last 15% of the interior of the sun, heat is transported by convection motions to just below the photosphere where radiation takes over again. Radiation in the form of light travels outward without interference of the chromosphere and corona. Acoustic energy generated in the convection zone heats up the chromosphere but mainly the corona.

are heated. Everything is bubbling and circulating. Just below the photosphere, neutral hydrogen atoms H and ions H$^-$ cause the zone to be opaque to radiation. It is for this reason that the photosphere "skin" energy flows again by radiation. Photons then travel on, practically unrestricted, as the solar atmosphere above the photosphere is mainly rarefied gas.

Directly above the photosphere is the chromosphere or sphere of color, which extends 10,000 km over the photosphere and reaches a temperature of 20,000 K. The shell is called the chromosphere because it has a light pink color. The chromosphere is visible during total eclipses of the sun. It can also be seen with a spectrohelioscope concentrated on wavelength 6563Å or H$_\alpha$ line. Hot jet gases called spicules (Figure 23.4) move matter and energy from the photosphere

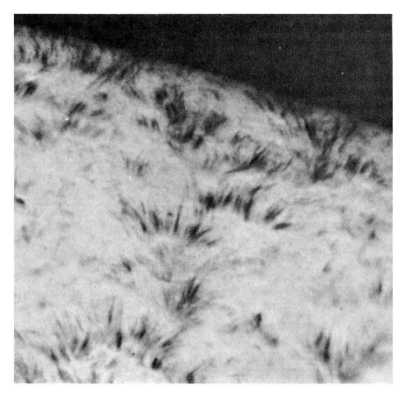

Figure 23.4. Hot jets of gas carrying matter and energy from the photosphere through the chromosphere to the corona. (California Institute of Technology)

to the corona. Above the chromosphere and extending over thousands and even millions of kilometers is the final layer of the sun called the corona, also seen during an eclipse. The corona appears as a faint halo of fine rays radiating from the sun. The shape of the corona varies. It is shown during an eclipse in Figure 23.5.

The sun rotates on its axis but not uniformly because of its gaseous consistency. At the equator, the average rotational period is 24.65 days. Halfway between the equator and the poles the period is 27.5 days, and near the poles 34 days.

The differential rotation of the sun is responsible, together with the sun's magnetic field, for several phenomena. It is obvious that the sun as a hot bowl of bubbling gas of various temperatures (20 million K at the center, 6000 K at the photosphere, and 2 million K in the corona), exposed to enormous gravitational forces that try to implode the star and the counteracting outward thermal pressure

The Fate of the Sun and Planetary System

Figure 23.5. The sun's corona shown during an eclipse. (Lick Observatory)

generated by the thermonuclear reaction in the core, is the scene of extremely interesting phenomena. We can cover only the highlights.

The sun's magnetic field, similar in strength to the magnetic field of the earth, becomes heavily wrapped around the sun because of the uneven rotation of the star. Often because of the twisting action, the magnetic field intensifies at certain areas and becomes up to 200,000 times stronger than the overall sun's magnetic field. When this occurs, the magnetic field ruptures the surface of the photosphere. This reduces the temperature in the area of rupture and reduces the emission and intensity of light. The spots which become darker are called sunspots (Figure 23.6).

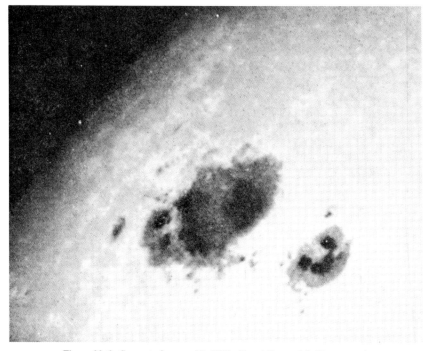

Figure 23.6. Sunspots January 20, 1920. (Royal Greenwich Observatory)

23.4. THE ACTIVITY OF THE SUN

The sun will remain approximately 5.5 billion more years on the main sequence until all of the hydrogen gas in the core is used up. During this period, however, its luminosity will gradually increase (Figures 23.7 and 23.8).

As a result of the previous 4.5 billion years of burning 600 million tons of hydrogen into helium each second, the sun's radiation has been supporting life on earth.

The hot corona contains electrons and protons. The high temperature, comparatively low gravity above the photosphere, and continuous movement of the corona cause the electrons and protons to escape from the sun in the form of the solar wind, which hits the earth at a high speed (about 500 km/sec). Huge solar eruptions (Figure 23.9) send intensely hot rarefied gases into the corona.

The core of the sun has a temperature of 15 to 20 million K during the hydrogen burning and the temperature decreases in the photosphere to 6000 K. But the temperature in the chromosphere rises to 10–20,000 K and to 2 million K

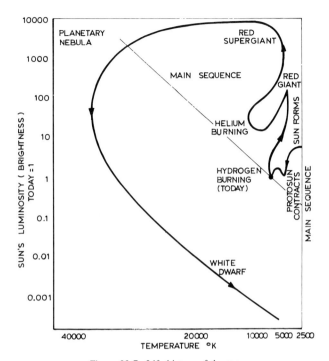

Figure 23.7. Life history of the sun.

in the corona. This phenomenon is explained as follows. In the convection zone just below the photosphere, the continuous turbulence and "boiling" of rising and sinking gas columns is extremely noisy and the created energy in the form of sound waves is responsible for the intensive heating of the chromosphere and mainly of the corona. The solar activity is substantial.

The number of sunspots and their intensity varies periodically in an approximate cycle of 11 years. Above a large sunspot there usually appears a larger disturbed region in the chromosphere called a plage. Such a plage can last several months. They often explode as flares and extend into corona seen as prominences which rise often thousands of kilometers above the photosphere (Figure 23.10). They are nearly always connected with sunspots and contain hot hydrogen gas.

23.5. THE FATE OF THE SUN

In 500 million years, when the sun's luminosity increases by 10%, the temperature will rise substantially. All water on earth will evaporate, CO_2 will be

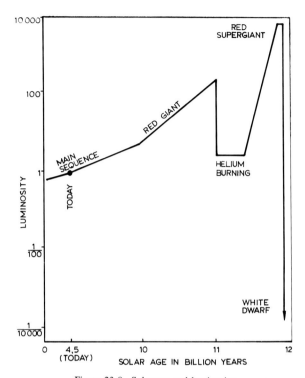

Figure 23.8. Solar age and luminosity.

released from rocks, and the climate will turn extremely hot, affected by a profound greenhouse effect. Earth will become the Venus of today and all life will cease.

In 1 billion years, the sun's luminosity will increase by 20%. Liquid water will appear on Mars.

In 3 billion years, the luminosity will increase by 50% and the climate on Mars will become earthlike with average temperatures of 25°C warmer than now.

In 5.5 billion years, all hydrogen in the central core of the sun will fuse into helium. The core will start to contract under gravitational forces and heat up. The increased temperature will trigger hydrogen burning in the shell above the core and radiation pressure from the shell will force the layers above to expand and cool. The sun will become a red giant, its expanded corona will cool down to 4000 K and glow red.

In 6 billion years, the sun's radius will increase to 0.2 AU or 30 million km, close to the planet Mercury; the sun will become a red giant, 50 times larger in

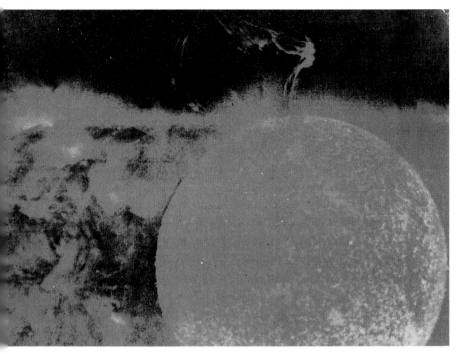

Figure 23.9. A huge solar eruption sends intensely hot rarefied gases over 300,000 km into the corona. Photographed from the orbiting space station Skylab. (NASA–Skylab)

the sky. The increased surface will cause the luminosity to increase 300 times. The surface temperature on earth will reach 800°C and the surfaces of Mercury and Venus will melt. The outer planets will thaw and undergo substantial changes.

In 7 billion years, as the core of the sun continues to contract and considerably heat up, the red giant phase will suddenly end. The temperature of the core will reach in a flash 100 million K, igniting its contents—helium—and the thermonuclear fusion of helium into carbon and oxygen will start (Figure 23.11). This stage will last approximately 100 million years.

Helium-4 collides with helium-4 forming beryllium-8, which is unstable. However, a further interaction with helium-4 creates carbon-12. Another collision of an atom of carbon-12 with helium-4 forms oxygen-16. During this period called the helium flash, the sun's luminosity drops to 60 from 300, the radius decreases, and its surface becomes hotter.

At the end of the helium-burning cycle, the sun's core will become carbon

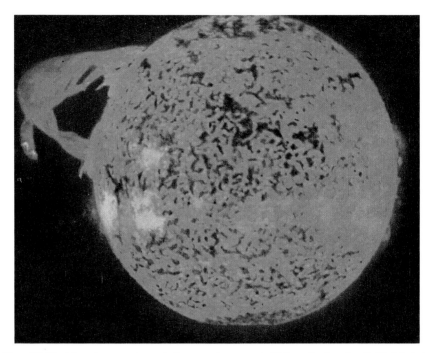

Figure 23.10. A loop-type prominence in the corona has grown to 40 times the size of earth. Skylab photo taken in ionized helium at 304 Å by NASA and NRL in 1973. (NASA–Skylab)

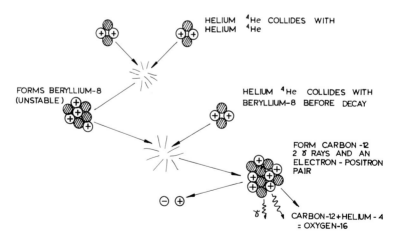

Figure 23.11. Creation of carbon and oxygen core.

and oxygen. Soon after, the core will start to contract and the temperature will also increase in the shells above. As a consequence, helium burning will be initiated at 100 million K in the shell and, next to it, hydrogen burning at 20 million K, as shown in Figure 23.12.

When the double shell burning commences, the sun's luminosity increases 10,000 times, the outer layers are pushed by the enormous radiation pressure from the two burning shells, the radius increases to 1 AU [from 7 million km (red giant) to 150 million km, or more than 20 times], and the sun becomes a red supergiant, as shown in Figure 23.13. The temperature in the solar system will increase tenfold. Mercury and Venus will be engulfed by the expanding sun and spiral inward and vaporize. The earth covered deeply with molten rock may survive moving to a different, more distant orbit. More than half of the sun's mass would be ejected, creating a planetary nebula. The temperature of all of the surviving planets including earth would go down considerably. The liquid rocks on earth

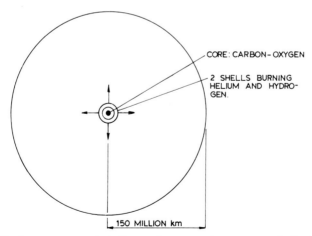

Figure 23.12. Star becomes a supergiant after core becomes carbon–oxygen and adjacent cells burn helium (100 million K) and hydrogen (20 million K).

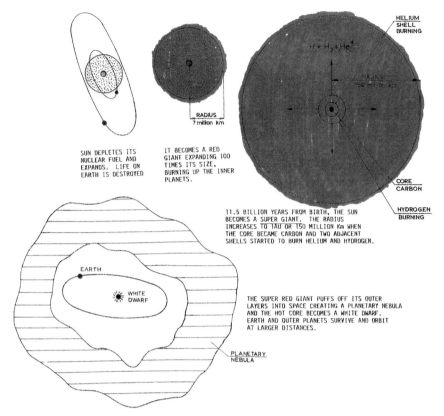

Figure 23.13. The fate of the sun 500,000 years from now to the creation of a white dwarf 5 billions years later.

will solidify again and the earth will become a dead and cold planet, orbiting the dim, small sun which has become a white dwarf, with considerably reduced luminosity.

24

The Death of Stars

24.1. THE DEATH OF THE SUN

In low-mass stars like our sun and up to 4 solar masses, when all helium is burned in the core into carbon and oxygen, burning of the shell helium and hydrogen signals the closely approaching death of the star, now a supergiant. After a cycle of contraction and reexpansion, the surface gravity of the supergiant will be so low that all of the outer layers will move out into space, creating a so-called planetary nebula around the now-naked, exposed very hot core. When all nuclear fuel is exhausted and fusion stops, the core will start again to collapse. The gravitational collapse of the core will stop at a density of approximately 1000 tons per cubic inch. Electrons, which are highly compressed at this density, resist all further compression and form what is called in physics degenerate electron matter. Degenerate matter, where electrons start to repel each other, is so resistant to further compression that nothing more can happen to the remaining naked core of the sun.

The sun, which by now will have shrunk to the size of earth or 10,000 miles in diameter with a surface temperature of approximately 120,000 C, will become a *white dwarf* containing the burned-up sun's matter in the form of carbon and oxygen. The planetary nebula will rapidly expand and thin out. The dramatic radiation of the nebula will dissipate after 40–60,000 years. White dwarfs cool off by radiating energy into space, mainly in the form of ultraviolet radiation, but the process is extremely slow.

What will be left of the sun is a lump of matter squeezed to the size of earth which will ultimately cool to a surface temperature of 5–6000 K. The luminosity will diminish to perhaps 1/10,000 of the sun's original brightness because of the small size. Even at 40,000 K the luminosity will be only 1/100 of its original level. The white dwarf, still glowing, will become a cold black dwarf and in the blackness of space the now long lifeless earth will pass into an eternal deep freeze.

There are billions of white and black dwarfs in the universe. About 600 white dwarfs have been sighted. A good example is shown in Figure 22.5 in the center of the Ring Nebula in Lyra (M57) or in NGC 6781 (Figure 24.1).

The resistance to gravitational collapse of remnants of small stars, or rather their cores, is provided by the so-called electron degenerate pressure. When the

Figure 24.1. White dwarfs are very common objects and are usually central stars of a planetary nebula. This is the nebula NGC 6781. (California Institute of Technology)

density of the core reaches 1000 tons/cm³, the electrons resist further compression and provide sufficient pressure to resist further implosion. Based on the Pauli exclusion principle of quantum mechanics, there is a minimum space in which electrons can be squeezed after which they vigorously resist any further compression. While the burned-out naked core of a star not larger than 1.44 solar masses becomes a white dwarf and is hot, the internal pressure resisting implosion is larger than the gravitational forces. Whether the white dwarf will remain in equilibrium after its temperature drops to absolute zero will depend on the resistance of the squeezed electrons counteracting gravitational contraction. Based on the Pauli exclusion principle, only a maximum of two electrons can have the same momentum $p = 0$. All others must have higher momentum which depends on the density of the electrons d_e.

The maximum Fermi momentum p_F is

$$p_F = 3.28 \times 10^{-27} d^{1/3} \tag{24.1}$$

The maximum energy E_F of the electron gas will for relativistic conditions be

$$E_F = c \times p_F = 0.60 \times 10^{-10} d^{1/3} \quad \text{(MeV)} \tag{24.2}$$

The gas pressure P_g is proportional to the density of the particles in the white dwarf d_e and the Energy E_F ($P_g \sim d_e E_F$):

$$P_g = 1.2 \times 10^9 \left(\frac{d}{K}\right)^{4/3} \quad \text{(erg/cm}^3\text{)} \tag{24.3}$$

where K is the number of nucleons for 1 electron. In a white dwarf, $K = 2$ (1 proton + 1 electron). (As an example: for a density $d = 10^7$ g/cm³, the internal pressure is 7×10^{10} kg/cm².)

The gravitational pressure P_{grav} depends on the mass M and radius R

$$P_{grav} \propto \frac{M^4}{R^4} \tag{24.4}$$

and in terms of the density d of the white dwarf

$$P_{grav} \propto M^{2/3} \times d^{4/3} \tag{24.5}$$

Comparing Eq. (24.5) with Eq. (24.3), it can be established that the pressure P_g can stabilize the star for a certain density d.

There is a critical mass of the white dwarf called the Chandrasekhar limit above which the star's electron degenerate pressure could not resist the gravitational pressure. It is:

$$M_{cr} \propto \frac{5.75}{K^2} M_\odot \tag{24.6}$$

where $K = 2$, two nucleons to one electron

$$M_{cr} \propto 1.44 M_\odot \qquad (24.7)$$

This is the limit at which a remnant of a star of up to 3 solar masses can survive gravitational collapse and become a white dwarf.

24.2. SUPERNOVA EXPLOSION OF A WHITE DWARF

When conditions described in the previous chapter become more radical, the entire white dwarf can explode and disintegrate. When an outer shell of gravitationally attracted hydrogen gas from a companion red giant star ignites thermonuclear burning and erupts, only a part of the accumulated shell ejects and the white dwarf remains intact. However, when the accumulation of hydrogen gas from a companion binary red giant star is a great deal larger and the total mass of the white dwarf increases substantially, more dramatic events can occur.

Because of the large mass accumulated on the white dwarf, the gravitational pressure increases greatly and the entire star starts to implode. The carbon–oxygen core of the star, surrounded by electron gas, heats up considerably. Eventually, when the temperature reaches 600 million K, the carbon ignites at the center. A shock wave develops, traveling outwards, destroying the white dwarf in a nuclear explosion called Supernova 1, ejecting a mass of 0.3–1 $M_\odot$. The binding energy, which kept the star in equilibrium, of 5×10^{50} erg becomes the kinetic energy of the ejected mass propelled with a large velocity of about 10,000 km/sec.

As we will see later, more powerful eruptions called Supernova 2 take place when the remaining mass of a dead star is larger than 1.44 solar masses and is subject to gravitational energy which is then released.

24.3. NOVA EXPLOSION OF A WHITE DWARF

Solitary white dwarfs of less than 1.4 solar masses can remain stable indefinitely. However, white dwarfs as companions of a binary star system can become a source of a dramatic nova explosion under the proper circumstances, as shown schematically in Figure 24.2.

The hydrogen gas of the companion red giant star is pulled by the intensive gravity of the hot and compact white dwarf. It accumulates on the hot surface and is heated up intensively by gravitational forces. When pressure increases enormously and temperature reaches the igniting level of 4 million K, the accumulated surface layer of hydrogen ignites and thermonuclear burning of hydrogen into

The Death of Stars

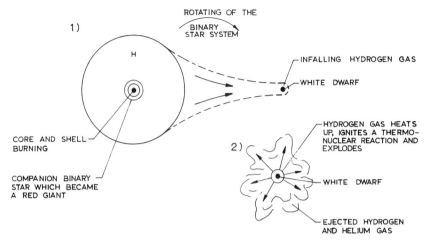

Figure 24.2. (1) Hydrogen gas of the companion red giant star is pulled toward the surface of the hot compact white dwarf. (2) The hydrogen gas is heated to 4 million K, ignites, and burns in a thermonuclear process. Luminosity increases 20,000-fold.

helium is initiated on the hot surface of the white dwarf. The overall mass of the white dwarf remains below the critical mass of 1.44 solar masses and the electron degenerate gas can still resist gravitation. Suddenly the luminosity increases up to 20,000 times and the white dwarf becomes a nova as seen in Figure 24.3. The burning hydrogen and helium are ejected into space. No more than 0.001 $M_\odot$ is ejected. After 40–70 days, the nova fades away, the white dwarf returns to its original conditions, and the ejected material forms a small glowing nebula, which also fades away soon after.

24.4. THERMONUCLEAR BURNING IN STARS 4 SOLAR MASSES

In stars larger than 4 suns, core collapse eventually ignites carbon burning at 600 million K, while simultaneously helium shell and hydrogen shell burning occurs. Carbon burning results in a star core containing oxygen, sodium, magnesium, and neon. Eventually, all of the carbon core burns out, the star core contracts and ignites the shell carbon helium and hydrogen burning (Figure 24.4).

In stars of a minimum 6 solar masses, temperature of the core after further contraction rises to 1 billion K and oxygen burning is initiated, resulting in a core of silicon and sulfur. When all oxygen is burned out, the core contracts,

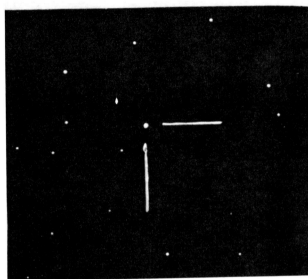

Figure 24.3. Nova explosion of a white dwarf surface shell of burning hydrogen in a binary system as seen from earth (left). After a few months the white dwarf returns to its original conditions (right). (Lick Observatory)

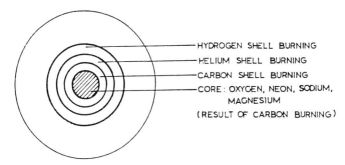

Figure 24.4. Schematic diagram of thermonuclear burning, stars 4–6 $M_\odot$.

temperature increases, and shell oxygen, carbon, helium, and hydrogen burning is initiated (Figure 24.5).

In stars larger than 8 sun, the process of thermonuclear reaction in the core and adjacent shells continues when the core temperature reaches 2 billion K and silicon burning is initiated, which results in an iron core with nickel. Iron nuclei are the most tightly bound nuclei in nature and do not burn. However, the temperature is sufficiently high to trigger thermonuclear burning in adjacent shells of silicon, oxygen, carbon, helium and hydrogen (Figure 24.6).

24.5. THE FATE OF A TYPICAL STAR, 18 $M_\odot$

24.5.1. The Birth of the Protostar

The mass of the sun $M_\odot$ is 2×10^{33} g, so we are speaking about the birth of a star with a mass of 36×10^{33} g and a radius of 30×10^{11} cm or 30 million km

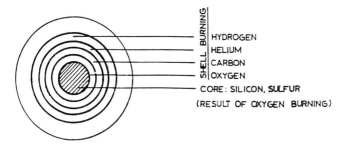

Figure 24.5. Schematic diagram of thermonuclear burning, stars 6–8 $M_\odot$.

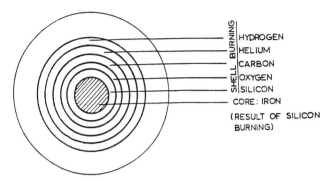

Figure 24.6. Schematic diagram of thermonuclear burning, stars $> 8\ M_\odot$.

when finally formed. The sun's present radius for comparison is 7×10^{10} cm, 700,000 km or 43 times smaller.

Stars condense from protostars, which form from interstellar or galactic gas and dust clouds due to self-gravity. The first protostars formed from pure primordial hydrogen and helium from which galaxies were created. These first protostars are called third generation. Later on, when the third-generation stars exploded in supernovas, the created planetary and emission nebulas contained, in addition to hydrogen and helium, heavier elements which were produced in thermonuclear reactions, mainly in the cores of large stars. And so, later on, second- and first-generation stars formed from nebulas and were enriched with heavier elements.

In most cases, self-gravity first pulls together large masses of interstellar gas and dust to form a large protocloud for a star cluster from which, later on, individual stars form. Typical clusters of young and old stars are shown in Figures 24.7 and 24.8.

The origin of a protostar of 18 solar masses is a lump of piled-up gas and dust that, by gravitational attraction, grew larger and larger when its mass and gravity was rising. And so, what could have been a small lump originally grew by gravitational accretion to an 18-solar-mass gas cloud that, under the effect of its own gravitational forces, started to implode into a formation often called a globule, as shown in Figure 24.9. As soon as it is formed, the globule begins to contract as it is unable to support the enormous masses of gas and dust pressing toward the center under its own gravity.

As the gravitational forces are most effective in the center, a core of higher density is being developed. The kinetic and gravitational energy of the infalling particles are converted into thermal energy. The thermal energy of particles depends directly on the number of particles and temperature, which soon rises

The Death of Stars

Figure 24.7. A cluster of young stars. (Lick Observatory)

Figure 24.8. A cluster of old stars. (Lick Observatory)

The Death of Stars

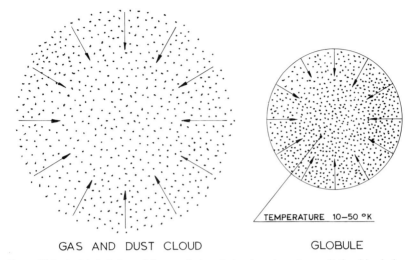

GAS AND DUST CLOUD   GLOBULE

Figure 24.9. A globule is formed from a piled-up dust and gas lump by gravitational implosion.

to as high as 400,000 K in the core where the density reaches a high level, and the contraction of the core slows down. The more particles per cubic centimeter, the higher is the thermal energy E_{th}:

$$E_{th} = 3NkT/2 \quad \text{and} \quad P = 2E/3$$

where N is the number of particles (protons, neutrons, electrons), k is the coefficient $= 1.38 \times 10^{-16}$, and P is the pressure. The internal gas pressure inside the core amounts to two-thirds the thermal energy. The contraction of the core stops for a moment when the internal gas pressure is sufficiently high enough to support the weight of the infalling outer layers. Once the core was created, which stabilized the total mass of the globule, a hydrostatic equilibrium was established and this event determines the birth of a protostar as shown in Figure 24.10.

As there is a substantially higher temperature in the core than at the surface, convection currents of hot material move outward while cold materials sink toward the core. After a time a thermal balance is established in addition to the hydrostatic equilibrium. However, as thermal energy is moving outward from the core, the internal pressure drops while the temperature of the core is lowered substantially.

The protostar is subjected to continuous contraction. Densities increase substantially in the core and adjacent layers of gas and heat are carried outward by radiation.

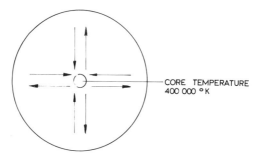

Figure 24.10. The creation of a protostar from a globule of 18 solar masses. Heat circulates by convection to establish a thermal equilibrium.

24.5.2. A Star Is Born

Temperature, density, and pressure increase continually due to the enormous gravitational implosion, which is equal to

$$E_G \cong G\frac{M^2}{R}$$

where M is the mass, R is the radius, and G is the gravitational constant.

The radius of the imploding star was reduced to 30 million km when the temperature of the core rose to 4 million K. At this level, the ignition temperature was reached and the core of the protostar of approximately 30% of the total was ignited to fuse hydrogen nuclei into helium.

24.5.3. Hydrogen Burning

For the next 10 million years, millions of tons of hydrogen nuclei fused into helium every second providing the necessary high temperature and pressure in the core to avoid collapse. The radius of the central core was 500,000 km and the mass 6.1 $M_\odot$.

Because of the intense thermonuclear burning and the large surface of the star, its luminosity at the time of hydrogen burning was about 40,000 times brighter than that of our sun. The final result of the hydrogen cycle reactions is the transmutation of four protons into a helium isotope $^4_2$He with an energy of 26.7 MeV liberated during the transmutation. Approximately 80% of the released energy is carried away by radiation of γ photons and 20% by neutrinos.

The following reactions occur in the hydrogen cycle:

$$p + p \rightarrow d + e^+ + \nu_e$$

The Death of Stars

$$d + p \rightarrow {}^3_2He + \gamma$$
$$ {}^3_2He + {}^3_2He \rightarrow {}^4He + 2p$$

If the ${}^3_2He$ vanishes in reaction, this puts an end to the chain:

$$4 \text{ protons} \rightarrow {}^4_2He + \text{energy}$$

The hydrogen nuclei were so close in the dense core and moved so fast that the temperature rose to 20 million K and triggered a thermonuclear reaction. Contraction stopped. The pressure balancing off the gravitation is provided by thermonuclear fusion. The ignition called hydrogen burning signaled *the birth of the star* and arrival on the main sequence (Figure 24.11).

24.5.4. Hydrogen Burning Ends

After the hydrogen burning cycle ended, the core of approximately 6.1 $M_\odot$ began a gradual contraction and the star underwent dramatic changes. During the

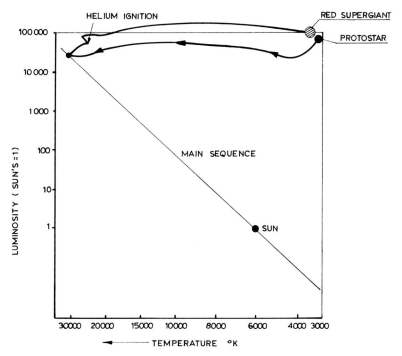

Figure 24.11. Evolution of the 18-solar-mass star before arrival on the main sequence and after the hydrogen-burning cycle ended and the star became a red supergiant.

hydrogen-burning cycle, the star's stability relied for 10 million years on thermal energy generated by the thermonuclear reaction. Now, after the burning stopped, the core could not support itself and imploded under the overwhelming influence of gravity. The inactive core was compressed by gravity and the density rose. At the same time, large amounts of gravitational energy were converted into heat, considerably increasing the temperature of the core.

24.5.5. Star Becomes a Red Supergiant

As soon as the temperature in the core raises the temperature in the layers adjacent to the core, a shell of hydrogen ignites (ignition temperature is 4 million K) and initiates hydrogen burning in the shell, while the thermonuclear inactive core continues to contract and heats up.

The new source of thermonuclear energy and the massive outpouring of radiation from the hydrogen burning in the shell pushes the star's outer layers, mostly hydrogen, causing an enormous expansion. The star becomes a red supergiant with a radius of 300 million km or 10 times the radius of the star during the hydrogen core burning.

24.5.6. Helium Ignition

After perhaps 30,000 years of constant contraction, the density of the core increases from 6 g/cm^3 to 1.1×10^3 g/cm^3. When the temperature of the core rises to over 100 million K, helium ignites and a new thermonuclear reaction is initiated. Soon the temperature rises to the helium-burning operating temperature of 190–200 million K, while the hydrogen shell burning continues. Schematically, this is shown in Figure 24.12.

As soon as the hydrogen shell burning is ignited, the star rapidly expands and the unburned hydrogen and helium gas cools down on the surface to approximately 4000 K, regardless of whether the star originally had 1 or 18 solar masses. However, as the star's surface expands enormously, the luminosity rises from 40,000 to perhaps 70,000.

As soon as helium ignites in the core, the brightness rises even more and may reach 105,000 of the luminosity of the sun. The reactions taking place during the helium-burning cycle are as follows. First, two helium-4 particles (or α particles) combine for a very short time ($\sim 10^{-16}$ sec) in an unstable beryllium $^8_4$Be isotope nucleus:

$$^4_2\text{He} + ^4_2\text{He} \rightarrow ^8_4\text{Be}$$

In spite of the very short lifetime of $^8_4$Be, due to high density it combines with another α particle and forms carbon:

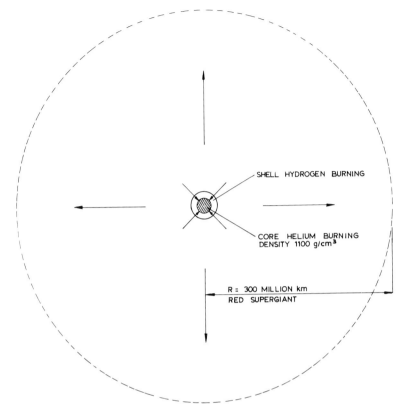

Figure 24.12. Helium burning. Star becomes a red supergiant with a radius of 300 million km.

$$^4_2He + {}^8_4Be \rightarrow {}^{12}_6C + \gamma$$

The effective thermonuclear reaction of burning helium is

$$3\,{}^4_2He \rightarrow {}^{12}_6C + 7.65 \text{ MeV}$$

After a large amount of carbon has been formed, a part of the helium will be consumed in further reaction of carbon with helium, creating oxygen.

$$^{12}_6C + {}^4_2He \rightarrow {}^{16}_8O + \gamma$$

There is also a possibility that small amounts of neon and magnesium are formed in the helium thermonuclear reactions:

$$^{16}_8O + {}^4_2He \rightarrow {}^{20}_{10}Ne + \gamma$$
$$^{20}_{10}Ne + {}^4_2He \rightarrow {}^{24}_{12}Mg + \gamma$$

After the helium-burning cycle is exhausted, lasting approximately 1 million years, the core of the star is made up from the carbon isotope $^{12}_{6}C$. It also contains the oxygen isotope $^{16}_{8}O$ and a small amount of $^{20}_{10}Ne$ and $^{24}_{12}Mg$.

24.5.7. The Next 12,000 Years of Thermonuclear Burning

As the star is sufficiently large to extend additional gravitational effects to squeeze the core, the process of contraction after a cycle of thermonuclear burning, consequent heating up of the core and ignition of the ash of the previous cycle of fusion continues for the next 12,000 years.

Carbon Burning

During the gravitational contraction after the helium cycle, densities substantially increased reaching 2.4×10^5 g/cm^3, raising the core's temperature. Once the core's temperature during the contraction reaches 600 million K, carbon burning is ignited. The process runs at an operating temperature of 740–800 million K. The following thermonuclear reactions are triggered during the combustion:

$$^{12}_{6}C + ^{12}_{6}C \rightarrow ^{20}_{10}Ne + ^{4}_{2}He$$
$$\rightarrow ^{24}_{12}Mg + \gamma$$
$$\rightarrow ^{23}_{11}Na + p$$

Oxygen also reacts with carbon:

$$^{16}_{8}O + ^{12}_{6}C \rightarrow ^{24}_{12}Mg + ^{4}_{2}He$$

$^{20}_{10}Ne$ is produced mainly with $^{24}_{12}Mg$ and $^{23}_{11}Na$.

Neon Burning

Once the core contracted again after the half of carbon burning, the density increased to 7.4×10^6 g/mm^3 and the temperature to 1.2×10^9 K, the burning of neon was triggered.

Neon $^{20}_{10}Ne$ can be considered to be oxygen $^{16}_{8}O$ bound to a helium nucleus $^{4}_{2}He$.

By the time neon started to burn at 1.2–1.4×10^9 K, the energy of radiation in the form of γ photons and neutrinos in the core was enormous. The energy of radiation, as we know, is directly proportional to the fourth power of the temperature: $E = aT^4$ (erg/K^4). The high-energy photons knock out a helium nucleus $^{4}_{2}He$ out of neon, creating oxygen:

$$^{20}_{10}Ne - ^{4}_{2}He \rightarrow ^{16}_{8}O - \gamma \quad \text{or} \quad ^{20}_{10}Ne + \gamma \rightarrow ^{16}_{8}O + ^{4}_{2}He$$

During this cycle, some silicon $^{28}_{14}Si$ is created in the following reactions:

$$^{20}_{10}Ne + ^{4}_{2}He \rightarrow ^{24}_{12}Mg - \gamma$$
$$^{24}_{12}Mg \rightarrow ^{4}_{2}He \rightarrow ^{28}_{14}Si - \gamma$$

The core after the burning cycle of neon consists mainly of oxygen with small quantities of silicon. As there are many neutrons in the core released by the reactions, they penetrate during collisions, not affected by electrostatic repulsion, into the nucleus creating new isotopes. Silicon can be converted into chlorine-35, and magnesium-26 into aluminum-27. The neon-burning cycle lasts only a short 12 days.

Oxygen Burning

When the neon-burning cycle ended, the star appeared already as a burning onion. While the core still contracted, the adjacent shell's layers burned neon, carbon, helium, and hydrogen, all at different temperatures. There is some mixing of elements between the core and the adjacent shells.

At this stage, when the density due to contraction of the core reached 1.6×10^6 g/cm^3 and the temperature level was raised to between 1.5 and 2.1×10^9 K, oxygen fusion was ignited and in a short period of 4 days the oxygen was transformed to silicon, phosphorus, potassium, sulfur, chlorine, calcium, and titanium. One of the fusion reactions is

$$^{16}_{8}O + ^{16}_{8}O \rightarrow ^{31}_{16}S + n$$

Silicon and Sulfur Burning

The ash of oxygen burning is silicon and sulfur. The thermonuclear fusion is halted and oxygen fusion is moved to the adjacent shell. When further implosion of the core takes place, the temperature in the core reaches 2.2×10^9 K and silicon burning is ignited.

The core's density increases to 5×10^7 g/cm^3. The core's mass is $1.9\,M_\odot$ and its radius is 5000 km. An enormous contraction took place in the 12 million years of the star's entire life from the time nuclear fusion of hydrogen was ignited and the star arrived on the main sequence.

After a week of silicon burning at temperatures reaching 3.5×10^9 K, thermonuclear burning in the star's interior stops abruptly when iron becomes the ash of silicon and sulfur burning. An iron does not burn regardless of how hot the star's core becomes, the star came abruptly to the end of its life. The structure of the star at the time the iron core was formed is shown schematically in Figure 24.13. One of the reactions of silicon burning is

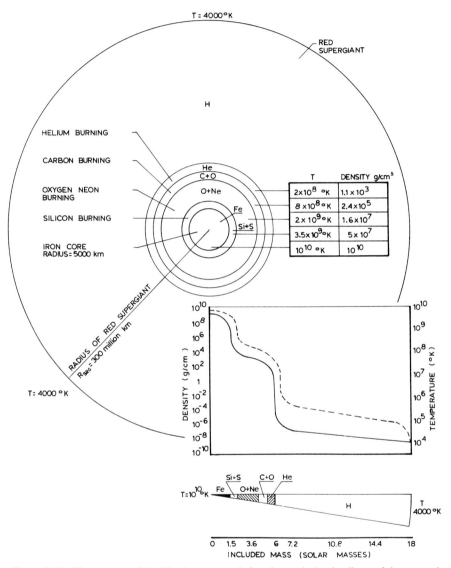

Figure 24.13. The structure of the 18-solar-mass star before the gravitational collapse of the core and supernova explosion. (Bottom diagram adapted from "How a Supernova Explodes" by Hans A. Bethe and Gerald Brown. Copyright *Scientific American*)

$$^{31}_{16}S + ^{31}_{16}S \rightarrow ^{56}_{26}Fe + 3^4_2He + \text{Energy } (\gamma)$$
$$3^4_2He \rightarrow 6p + 6n$$

24.5.8. The Collapse of the Core and Supernova Explosion

Figure 24.13 shows schematically the structure of the 18-solar-mass star at the time when silicon and sulfur, burning in the core, formed an iron core with a thin shell of silicon burning which is adding mass to the iron core. The star has at that moment an inert core of iron of approximately 1.5 solar masses and a very small radius of 5000 km. The fusion shell burning of oxygen, neon, carbon, silicon, and helium in adjacent shells continues in the inner 6 solar masses, while the enormous envelope of mostly hydrogen gas of 12 solar masses spread out in the red supergiant with a radius of 300 million km, continues to implode toward the core.

The $3.5\text{--}5 \times 10^9$ K temperature and enormous density of the core drop off substantially in the burning inner part of the star and then further drop substantially in the envelope. The red supergiant has a surface temperature of only 4000 K but sufficiently high to glow with a brightness of over 100,000 times the luminosity of the sun.

As all thermonuclear reactions in the core stopped, the hydrostatic equilibrium which was achieved temporarily, began to become unbalanced. The iron core which was built up in 1 week of burning silicon and sulfur will collapse in less than 1 sec.

We will now analyze in detail the anatomy of the core and the events that cause the core to collapse and bounce back, creating shock waves that result in an explosion of colossal proportions known as Supernova 2.

24.5.8.1. Stage 1—Breakup of Iron Nuclei

No atoms can survive under the extreme density, pressure, and temperature levels of the iron core. The density reaches 10^{10} g/cm^3 and the temperature 5×10^9 K. The core consists of ionized iron nuclei, floating in a mass of electrons.

The collapse begins when the mass of the iron core exceeds the Chandrasekhar limit of 1.4 solar masses. The pressure of the electrons created by their high energy due to the exceedingly high temperature of the core and their mutual electromagnetic repulsion can no longer resist gravitational implosion. In addition, the energy of the core is absorbed by the breakup of the iron nuclei (Figure 24.14).

The iron nuclei $^{56}_{26}Fe$ made up of 26 protons and 30 neutrons break up in collisions with high-energy photons first, to nuclei of helium ^4_2He and neutrons absorbing enormous amounts of thermal energy from the core in the process. The reactions that take place are

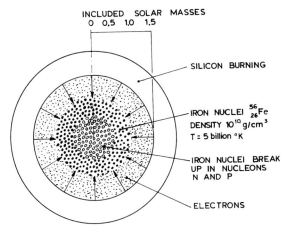

Figure 24.14. First stage of core collapse. Iron nuclei break up into protons and electrons, reducing energy level, pressure, and temperature.

$$\gamma + {}^{56}_{26}\text{Fe} \rightarrow 13{}^{4}_{2}\text{He} + 4n - 124 \text{ MeV}$$

$$\gamma + {}^{56}_{26}\text{Fe} \rightarrow 26p + 30n - 493 \text{ MeV}$$

The process is extremely fast. At prevailing temperatures of the core of 10^{10} K, it takes less than 10^{-6} sec. What happens is that in the time of less than a second, the energy absorbed during the iron nucleus breakup is equal to what the star has radiated during its entire active life through the thermonuclear fusion cycle. The energy comes mainly from the electrons, which considerably reduces their pressure, causing the collapse to increase.

24.5.8.2. Stage 2—Electron Capture

The center of the core at this stage consists of nucleons, protons, neutrons, and electrons. What takes place now due to enormous inward gravitational pressure is the so-called electron capture reaction. Electrons are virtually squeezed into protons to yield a neutron and neutrino:

$$p + e^- \rightarrow n + \nu_e$$

The neutrinos at this stage escape from the star carrying off energy and causing a cooling effect.

The second stage of the core collapse comes to an end when the density reaches 10^{11} g/cm^3 preventing further escape of neutrinos, which become trapped for a while anyway (Figure 24.15).

The Death of Stars

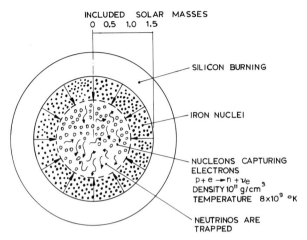

Figure 24.15. Second stage of core collapse. Nucleons capture electrons, pressure resistance reduced; implosion continues.

24.5.8.3. Stage 3—Core Reaches Nuclear Density

Due to the capture of electrons and reduced pressure, the gravitational collapse continues until the core reaches nuclear density which is approximately 2.7×10^{14} g/cm³. Neutrons and some protons and electrons form a continuous fluidlike medium known as nuclear matter. The initial repulsion or strong interaction and the high thermal energy of the particles is strong enough to balance off the gravitational forces, permitting the creation of a neutron star from the remnants of the core after the supernova explosion. At ground-state nuclear density, the distance between the nucleons is 1.8 fm [fm (fermi) = 10^{-13} cm] or 1.8×10^{-13} cm, which is just twice the radius of a proton (0.9×10^{-13} cm). At his ground stage state, the quark structure of the nucleons is not impeded.

24.5.8.4. Stage 4—Core Bounces, Shock Waves, Supernova 2 Explosion

Nuclear matter in the central part of the core has a high resistance to compressibility and is the major source of the creation of a shock wave during further implosion, which will cause the explosion (Figure 24.16). Once the center of the core reaches nuclear density, the outer shell material is still falling inward with high velocity, reducing the star's radius and increasing the gravitational implosion forces.

The central part of the core is squeezed to the maximum compression nuclear matter can take before collapsing itself. After the maximum squeeze, which is

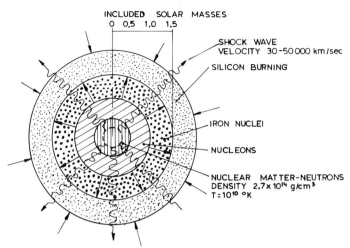

Figure 24.16. Last stages before Supernova 2 explosion. First the core is squeezed to nuclear density 2.7×10^{14} g/cm³. Subsequent enormous compression of the core to 4 times nuclear density creates a bounce back and powerful shock waves that explode the outer layers of the star.

estimated to reach approximately 4 times the ground-state nuclear density or 10.8×10^{14} g/cm³, the sphere of nuclear matter estimated now at 0.4 solar mass and a radius of 5 km only, bounces back like a rubber ball when squeezed, creating shock waves, resulting in a spectacular explosion of all outer material. The shock waves move with high speeds of 30–50,000 km/sec.

The shock explosion carries with it an enormous amount of kinetic energy estimated at 10^{55} erg and the ejected mass moves with a velocity of 5000 km/sec. The energy and matter carried outward seem to be sufficiently high to avoid being stalled by the mass of outer shell material, still falling inward.

Some scientists propose an alternative model of supernova explosion in two steps. The initial shock wave in this model stalls approximately at a radius of 150 km but is released again by reheating caused by the release of a shower of trapped neutrinos.

An estimated three or four supernovas explode in our own galaxy every 100 years. A typical remnant of a supernova explosion is shown in Figure 24.17, called the Crab Nebula. The star exploded in A.D. 1054 and was observed and recorded by the Chinese. It is about 8 light-years in diameter.

The remnants of the star of approximately 1.5 solar masses settle into a neutron star and liberate their trapped gravitational energy E_G. The formation of a neutron star with a mass $M = 1.5 M_\odot$ liberates energy of approximately

The Death of Stars

Figure 24.17. A visible neutron star in the Crab Nebula. The star exploded in A.D. 1054. (Lick Observatory)

$$E_G = \frac{GM_n^2}{R_n} = \frac{G(1.5\ M_\odot)^2}{R_n}$$

where G is the gravitational constant = 6.7×10^{-8}, M_n is the mass of the neutron star = $1.5 \times 2 \times 10^{33}$ g/cm^3, and R_n is the radius = 5×10^3 km:

$$E_G = \frac{6.7 \times 10^{-8}(3 \times 10^{33})^2}{5 \times 10^3} = \frac{6.7 \times 10^{-8} \times 9 \times 10^{66}}{5 \times 10^3} = 12 \times 10^{55}\ g$$

At the time of a supernova flash, the star's luminosity grows billions of times and for a short time the brightness of the star compares with that of an entire galaxy (Figure 24.18). A typical Supernova 2 explosion was observed in May 1987 of a supergiant star called Sanduleak–69°202. The star was about 80,000 times brighter than the sun but its luminosity reached, at the height of the explosion, 280 million times that of the sun. Sanduleak–69°202 was originally a star of approximately 18 solar masses and therefore reflects the schematic description of events in this chapter.

24.6. NEUTRON STARS, PULSARS

We have just described the fate of a typical large star of 18 solar masses. The remnant core of the star after the Supernova 2 explosion, which emitted into space most of the star's contents, became a highly compact neutron star with a diameter of roughly 20 km or 1/500 that of a white dwarf. While the density in the center of a white dwarf with an approximate diameter of 10,000 km is about 1000 tons/cm^3, the density of a neutron star in its center is a staggering 10 billion tons/cm^3.

We have seen that a white dwarf reaches equilibrium and stability against gravitational implosion by the resistance of electrons squeezed to the maximum density called an electron degenerate gas. They can resist gravitation exerted by a mass not exceeding, even in the cold state without thermal energy of particles and radiation, 1.44 solar masses. In a similar way as electrons, neutrons when squeezed to a so-called nuclear density of 2.7×10^{14} g/cm^3, resist further compression mainly due to substantial nuclear repulsion forces acting between neutrons at short range, as described in an earlier chapter on the behavior of nucleons, in addition to the Pauli exclusion principle, which does not allow more than two neutrons to have the same momentum of $p_n = 0$.

Though the core of a neutron star consists mainly of neutrons, there are some free electrons, and to equalize their negative electrical charge an equivalent number of protons with positive charger are present to make the neutron star electrically neutral.

At nuclear density of 2.7×10^{14} g/cm^3, there are only an estimated 1 electron

Figure 24.18. The Supernova 2 explosion of a supergiant star of approximately 18 solar masses with a luminosity of 80,000. During the height of the explosion of the star Sanduleak−69°202, its luminosity increased 200 million times. A few months later—a neutron star remnant shown in the inset. (Anglo-Australian Observatory)

and 1 proton for each 100 neutrons. The neutron degenerate gas can stabilize a star even at absolute zero temperature, provided its mass does not exceed 2.5 solar masses:

$$M_n < 1.5\text{–}2.5 \ M_\odot$$

Nearly all neutron stars being too small (20 km diameter) can be identified as pulsars. Pulsars are neutron stars which spin like giant tops at enormous speeds. They have strong magnetic fields which prevent the radiation of the hot neutron star from escaping except at the magnetic poles. For this reason, their radio signals sweep across space like beams from lighthouses (Figure 24.19).

The neutron star in the Crab Nebula (Figure 24.17), a remnant of a Supernova 2, was the first identified pulsar, at the National Radio Astronomy Observatory in West Virginia. It remains the fastest pulsar, emitting pulses about 30 times a second. The discovery of pulsars was made by Henish and Bell in Cambridge, U.K.

There are basically three phenomena responsible for the pulsating mode of a neutron star and the principal properties of pulsars. First is the high conductivity of the hot (near 100 billion K) plasma of a newly formed neutron star. Even within

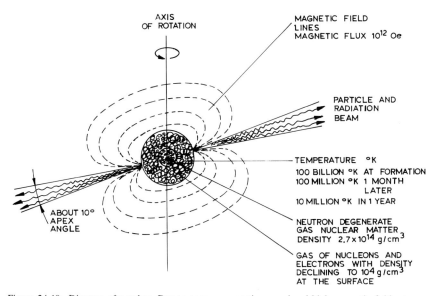

Figure 24.19. Diagram of a pulsar. Due to enormous rotation speed and high magnetic field, stars can radiate photons, neutrinos, and particles only within a narrow cone (10°) at the north and south magnetic poles.

1 month, the temperature will still be 100 million K and most of the heat energy is carried away by neutrinos. In 1 year, the temperature still remains at 100 million K. Because of the high conductivity, the strength of the magnetic field of the original star does not change during the gravitational implosion. However, even a small magnetic field spread over an enormous surface such as the 18-solar-mass star, analyzed in a previous section, would become highly compressed and intense when reduced to a tiny surface of the star remnant, in the form of a neutron star, with a diameter of 20 km.

The relationship between the magnetic field M_F and the radius of the star R must remain constant:

$$M_F \times R^2 = \text{constant}$$

The original radius of the 18-solar-mass star of 30 million km was reduced to 20 km, a proportion of 1.5×10^6. Consequently, the magnetic flux of the neutron star would increase by 2.25×10^{12}.

The intensity of the magnetic flux of an average pulsar is 10^{12} Oe ($M_F \sim 10^{12}$ Oe). As well, the magnetic axis usually does not coincide with the axis of rotation, as shown in Figure 24.19.

In addition, in accordance with the conservation law of the rotational momentum, the relationship between radius and rotation period must remain constant:

$$\frac{R^2}{T} = \text{constant}$$

where R is the radius of the star and T the period of rotation. Accordingly, for our 18-solar-mass star, the period of rotation must increase during the contraction to a neutron star by 2.25×10^{12} times. Typically, the period of rotation of neutron-pulsar stars is

$$T \approx 0.1\text{--}0.01 \text{ sec}$$

As a consequence of the high speed of rotation and the gigantic magnetic field, the radiation of photons, neutrinos, and particles can only take place within a small radiation cone of $10°$ at the north and south magnetic poles, as shown schematically in Figure 24.19.

The pulsed characteristic of the radio-frequency radiation from neutron pulsars is caused by the "lighthouse effect" when the earth enters the radiation cone (Figure 24.19). At the time of formation of the very hot neutron star with temperatures of 100 billion K, most radiation is transmitted at a wavelength of 0.03Å, corresponding to X rays at enormous power output. Later, when the star cools off, radio waves are emitted.

The 100 billion K hot matter generates thermal energy that, together with the resistance of the degenerate neutron gas, creates the balance against further implosion. The star then cools down mainly by the radiation of neutrinos, which carry off most of the energy. An estimated 3×10^{53} ergs equals nearly 100 times the energy carried by the supernova explosion or 10% of the total mass of the neutron star.

Electromagnetic radiation reflects mostly the events on the surface rather than the interior of the neutron star which after 1 year cools down to 100 million K or 1000-fold. In some cases a neutron star can, shortly after the supernova explosion, attract by gravitational pullback part of the material emitted into space by the supernova and increase its total weight beyond the threshold limit of a neutron star, to resist gravitation and undergo further implosion into a black–white hole, as discussed later.

Surface Nova Explosion on Neutron Stars in Binary Systems

Some neutron stars in binary systems display regular bursts, smaller in strength than nova explosions, which sometimes destroy white dwarfs in binary systems. A white dwarf in a binary system draws material by gravitational attraction from its companion star, usually in the form of a red giant. This will eventually trigger thermonuclear burning of the carbon core and a subsequent nuclear blowout of the entire star. In the case of neutron stars, the accumulation of the drawn-in material affects only the surface shell. A neutron star would capture from its companion red giant star yearly an estimated 10^{-9} $M_\odot$ material. The drawn-in gases hit the extremely hot, high-energy surface of a neutron star and heat up to temperatures that trigger explosive thermonuclear burning of helium at 100 million K. A short-lasting explosion takes place resulting in the release of intensive radiation of powerful X rays.

24.7. A NEW CONCEPT FOR BLACK HOLES

Present theories are dismissed in the section on gravity. When a remnant core of a large star after a Supernova 2 explosion exceeds 2.5 solar masses, the quantum internal pressure created by the resistance of particles of matter to further compression at nuclear density of 2.7×10^{14} g/cm^3 is not sufficient any more to oppose the overwhelming gravitational forces.

According to prevailing theories, such a core is soon, after all thermonuclear burning stopped, crushed in a rapid, total gravitational implosion lasting less than 1 sec, into a mathematical point with infinite density, temperature, and gravity, called a singularity. The space in the immediate vicinity of the singularity or black hole is totally warped.

The Death of Stars

The gravitational energy E_{GRAV} of a mass M and radius R established earlier is

$$E_{GRAV} \cong \frac{GM^2}{R}$$

We can see quite clearly from this equation that regardless of how large the mass M of the collapsing body is, when the radius R becomes 0, the gravitational energy becomes infinite ($E_{GRAV} = GM^2/0 = \infty$).

It is clear that the ultimate fate of gravitational energy materializes in a black hole. Always in accordance with the present theory, even before the collapsing star becomes a singularity, it reaches during the implosion a threshold size called the gravitational sphere or horizon. The horizon comes into being after the ultimate collapse into a black hole. Its radius, called the gravitational radius, depends entirely on the mass M of the collapsing star and is calculated from a previously established equation:

$$r_g = \frac{2GM}{c^2} \quad \text{(cm)}$$

Also, time slows down near the horizon and becomes 0 at singularity, wound together with space around the singularity. The remnants of the star become self-contained, completely isolated from observation, and, for all practical purposes, isolated from the universe.

Conventional black holes, even if they were to radiate, could not be seen due to their small size. For example, the gravitational radius of a black hole of 1 solar mass is only 6 km and for a large star with a core of 10 solar masses, the radius of the horizon is only 60 km.

The present theory has not eliminated the considerable difference in opinions regarding the contents of black holes. According to one version, it contains nothing except the warped space around it. Other black hole specialists claim that black holes contain the entire mass of the star remnants crushed to infinite density in zero space. Regardless, all of the experts agree that physical laws governing the universe, including quantum mechanics, break down and do not apply to a singularity.

When analyzing the processes taking place inside a rapidly imploding large core of a star exceeding 2.5 solar masses, one comes to interesting conclusions, differing from present theories. Let us analyze the step-by-step fate of a collapsing core of 3 solar masses following the Supernova 2 explosion of a large star.

First stage: Matter at nuclear density 2.7×10^{14} g/cm^3, radius 8.5 km, temperature 100 billion K.

We have analyzed the fate of an 18-solar-mass star in a previous section with the core becoming a neutron star. Here, just when all thermonuclear core and shell burning stopped, the naked, exposed remaining core contains 3 solar masses and is too large for settling down to a permanent fate of a neutron star. It has nuclear density of 2.7×10^{14} g/cm³, an approximate temperature of 100 billion K, and a radius of 8.5 km calculated below.

The radius R of the collapsing core at density $d = 2.7 \times 10^{14}$ g/cm³, having a mass of $3M_\odot = 6 \times 10^{33}$ g ($M_\odot = 2 \times 10^{33}$ g), can be calculated from the equation of volume versus mass and density:

$$\text{(Volume)} \quad \frac{4\pi R^3}{3} = \frac{M}{d} = \frac{6 \times 10^{33}}{2.7 \times 10^{14}} = 2.2 \times 10^{19} \text{ cm}^3$$

$$R^3 = 0.52 \times 10^{19}$$

$$R = (0.52 \times 10^{19})^{1/3} \sim 1.72 \times 10^6 \text{ cm} = 17.2 \text{ km}$$

$$\text{Diameter} = 2R = 34.4 \text{ km}$$

The nuclear matter at this stage consists mostly of neutrons and is a continuous, fluidlike medium. Nearly all electrons were squeezed into neutrons. The density of 2.7×10^{14} g/cm³, otherwise called ground density, corresponds to a numerical nucleon density of $d_0 = 0.15$ fm³ [1 fermi (fm) = 10^{-13} cm ~ radius of a proton or neutron] and an energy density of $E_0 = 0.14$ GeV fm^{-3}. The equivalent and prevailing temperature is 100 billion K.

The thermal forces created by the rapid movements and collisions of particles and radiation resist the imploding forces of gravitation only with limited success. The particles of matter of the star, mainly neutrons at this highly elevated temperature, collide with high energy and the generated heat creates a large thermal force, which can be calculated from the formula

$$E_{\text{THERM}} = 3(N) \times kT \quad \text{(erg)}$$

where N is the number of particles and k is the Boltzmann constant = 1.38×10^{-16} erg K. To calculate the thermal energy of the particles, we must first establish the number N of neutrons in the mass of 6×10^{33} g. The rest mass of a nucleon $m_n \cong 1.672 \times 10^{-24}$ g. The number of particles is

$$N = \frac{M}{m_n} = \frac{6 \times 10^{33}}{1.672 \times 10^{-24}} = 3.5 \times 10^{57} \text{ particles}$$

$$E_{\text{THERM}} = 3 \,(3.5 \times 10^{57}) \times 1.38 \times 10^{-16} \times 10^{11}$$

$$\sim 14.5 \times 10^{52} \text{ erg}$$

The gravitational force at radius $R_{\text{FB}} = 1.72 \times 10^6$ cm can be calculated from

$$E_{\text{GRAV}} \cong \frac{GM^2}{R_{\text{FB}}} \quad \text{(erg)}$$

where $G = 6.673 \times 10^{-8}$ cm³/g sec²

$$E_{GRAV} = \frac{6.673 \times 10^{-8}(6 \times 10^{33})^2}{1.72 \times 10^6}$$

$$= 140 \times \frac{10^{-8} \times 10^{66}}{10^6}$$

$$= 140 \times 10^{52} = 1.40 \times 10^{54} \text{ erg}$$

The calculated gravitational energy of 1.40×10^{54} erg is larger than the thermal energy of particles of 1.45×10^{51} by approximately 1000 times and the star implodes rapidly. The kinetic energy of the infalling particles $Mv^2/2$ with a speed nearing the speed of light is turned into thermal energy heating up the imploding star.

Second stage: Quark–gluon plasma at density 1.3×10^{15} g/cm, radius 8 km, temperature 10^{13} K.

The collapse continues with enormous velocity, increasing temperature and density. At approximately 10 times the ground state or nuclear density or 2.7×10^{15} g/cm^3 corresponding to a temperature of 1000 trillion K, quarks and gluons escape the strong nuclear force confinement within the nucleus and matter in the imploding star becomes for a short time a quark–gluon–electron plasma, similar to the medium in the Velan primordial fireball of the newly born universe.

Quarks and electrons are squeezed by gravity to a density of 2.7×10^{15} g/cm^3 and the temperature rises to 10^{13} K:

$$\text{(Volume)} \quad \frac{4\pi R^3}{3} = \frac{M}{d} = \frac{6 \times 10^{33}}{2.7 \times 10^{15}}$$

$$R^3 = 0.53 \times 10^{18}$$

$$R = (0.53 \times 10^{18})^{1/3} = 0.8 \times 10^6 \text{ cm} = 8 \text{ km}$$

The diameter of the star is reduced to only 16 km. At this high temperature level, the three forces of nature are unified and only gravitation is active.

Third state—Transmutation of matter into gamma energy: Temperature of matter 10^{15} K, density 2.7×10^{18} g/cm^3, radius 0.79 km.

The collapse proceeds rapidly and the radius is reduced to only 0.79 km. It is my opinion that at the corresponding density 10,000 times greater than nuclear density, not even the quarks and electrons can survive the squeeze and in a dramatic event all matter is suddenly transmutated into gamma energy in accordance with the Einstein equivalent formula $E_R = Mc^2$. This event takes less than 10^{-6} sec to materialize. Contrary to an immediate reverse procedure which occurs in high-energy particle accelerators, whereby new particles are created from the energy in a collapsing star to a "black hole," the gravitational forces are so overwhelming that the gamma energy remains intact.

Let us calculate the gamma energy which was formed from the transfer of matter into energy. To establish the energy of the transmutation of mass $M = 6$

$\times 10^{33}$ g, using $c = 3 \times 10^{10}$ cm/sec, we must convert grams, centimeters, and seconds into electron volts in which energy is measured. Using 1 g = 5.6×10^{32} eV, 1 cm = 5×10^4 eV, and 1 sec = 1.5×10^{15} eV,

$$E = M \times c^2 \quad (\text{g} \times \text{cm}^2/\text{sec}^2)$$

$$\text{Conversion constant } \frac{\text{g} \times \text{cm}^2}{\text{sec}^2} = \frac{5.6 \times 10^{32} \times 25 \times 10^8}{2.25 \times 10^{30}} = 62.2 \times 10^{10}$$

$$E_R = M \times (3 \times 10^{10})^2 \times 6.22 \times 10^{11}$$
$$= M \times 56 \times 10^{31} \text{ eV} \tag{24.8}$$

The mass $M = 3M_\odot$ or 6×10^{33} g was transmutated to gamma energy with the equivalent of

$$E_R = 6 \times 10^{33} \times 56 \times 10^{31}$$
$$= 3.36 \times 10^{66} \text{ eV}$$

The radiation of 3.36×10^{66} eV, though equivalent in energy to the transmutated mass of matter 6×10^{33} g, is contained in a much smaller volume than matter. The gamma density can be calculated from a now well-known equation:

$$d_r = aT^4 \quad (\text{erg/cm}^3)$$

1 erg = 6.25×10^{11} eV

$$d_r = aT^4 \times 6.25 \times 10^{11} \quad (\text{eV/cm}^3)$$

$a = 7.56 \times 10^{-15}$ erg/cm^3 K^4. In electron volts:

$$a_{\text{eV}} = 7.56 \times 10^{-15} \times 6.25 \times 10^{11} \text{ eV/cm}^3 \text{K}^4$$
$$d_r = 47.25 \times 10^{-4} \times T^4 \quad (\text{eV/cm}^3)$$

At $T = 10^{15}$ K

$$d_r = 47.25 \times 10^{-4} \times 10^{60}$$
$$= 4.725 \times 10^{57} \text{ eV/cm}^3$$

To find the volume we must divide the total energy in electron volts by the density d_r:

$$(\text{Volume}) \quad \frac{4\pi R^3}{3} = \frac{E_r}{d_r} = \frac{3.36 \times 10^{66}}{4.725 \times 10^{57}} = 0.71 \times 10^9 \text{ cm}^3$$

The radius of the imploding star R

$$R^3 = 0.166 \times 10^9$$
$$R = (0.166 \times 10^9)^{1/3} = 0.55 \times 10^3 \text{ cm} = 550 \text{ cm}$$

The Death of Stars

We can now calculate the forces or rather energy of gravitation and thermal forces of the imploding radiation at $T = 10^{15}$ K and $R = 550$ cm to see if they are in equilibrium.

Energy of gravitation is:

$$E_G = \frac{GM^2}{R} = \frac{6.673 \times 10^{-8} (6 \times 10^{33})^2}{550} = 0.44 \times 10^{58} \text{ erg}$$

Energy of radiation is:

$$E_R = aT^4 \times (\text{volume at } R = 550 \text{ cm}) \quad \text{erg}$$

The temperature of radiation dropped in inverse proportion to the scale of expansion of the universe during formation, but in reverse the temperature increases with the reduction of the imploding mass in the same relation.

At the time of transmutation of the entire matter of the collapsing core, the radius was 0.79×10^5 cm at $T = 10^{15}$ K.

At the instance of creation of the gamma energy, the gravitational forces squeezed the radiation to a much smaller volume of a radius of 550 cm from 790×10^3 cm or by a factor of 10^3.

The temperature of the radiation ball was therefore increased to $T = 10^{18}$ K:

$$E_R = aT^4 \times (\text{volume}) \quad \text{erg}$$
$$= 7.56 \times 10^{-15} (10^{18})^4 \times 0.77 \times 10^9 \text{ erg}$$
$$= 5.82 \times 10^{66} \text{ erg} \quad > \quad E_{Gr} = 4.5 \times 10^{57} \text{ erg}$$

At this stage, the radiation energy was one billion times more powerful than the energy of gravitation. Consequently, a hold was initiated in the rapid collapse of the star which momentarily expanded, radiated off part of the radiation, and cooled down. Soon after, however, as the gravitational energy of the original mass of 6×10^{33} g varies only with the radius and a cooler radiation cannot resist the gravitational forces, the implosion continued.

We have shown here a multistep implosion process that will finally stop when the gravitational energy is so strong that even thermal radiation cannot escape. We can calculate the radius of the equilibrium at the various stages and relationships between mass, radius, and temperature as follows.

If equilibrium exists, then gravitational forces must be equal to the thermal forces of radiation:

$$\frac{GM^2}{R} = aT^4 \times \left(\frac{M^4}{3}R^3\right)$$

$$GM^2 = aT^4 \times 4.1R^4$$

$$R = \left(\frac{GM^2}{a \times 4.1 \times T^4}\right)^{1/4} = \frac{3.75}{T} \times (M^2)^{1/4} \tag{24.9}$$

For our case of $M = 6 \times 10^{33}$

$$R = \frac{1}{T} \times 2.6 \times 10^{18} \text{ cm}$$

at $T = 10^{18}$ K

$$R = 2.6 \text{ cm}$$

In this theory we are getting finite results, eliminating the troublesome mathematical point or singularity. Even at Planck's temperature of 10^{32} K

$$R = \frac{2.6 \times 10^{18}}{10^{32}} = 2.6 \times 10^{-14} \text{ cm}$$

As most stars rotate, a black hole or the collapsed core of the star must rotate as well. However, as the rotational momentum must remain constant to meet the conservation law, the relationship between the final radius R_{BH} of the black hole and the period of rotation t must remain constant:

$$\frac{R_{\text{BH}}^2}{t_{\text{BH}}} = \text{constant} = \frac{R_0^2}{t_0}$$

Consequently, the black hole with the small radius spins like a top at enormous speeds.

In Figure 24.20 we compare the structure of a Velan rotating black hole of 3 solar masses with the classical Kerr black hole. In the Kerr rotating black hole the mathematical point singularity rotates forming a ring. The event horizon has a diameter of 18 km for the 3 solar masses. Temperature, density, and gravity are infinite. The contents of the black hole are not determined. In the Velan rotating black hole, the diameter of the black hole is 5.2 cm at an estimated temperature of 10^{18} K. It can vary depending on when the rapid implosion stopped due to inertia.

As shown in Figure 24.21, even at Planck's temperature of 10^{32} K it has a final diameter of 5.2×10^{-14} cm, slightly smaller than a proton. Temperature, density, and gravitational energy which curves the space around the black hole can be easily determined.

As the temperature of 10^{15} K has been reached when the quark–gluon–electron plasma annihilated into electromagnetic energy, it is estimated that the event horizon has been established when the imploding star reached a diameter

The Death of Stars

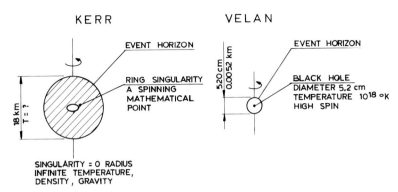

Figure 24.20. Comparative structure of a Kerr black hole with spin but no electrical charge and a Velan-type black hole, both having 3 solar masses.

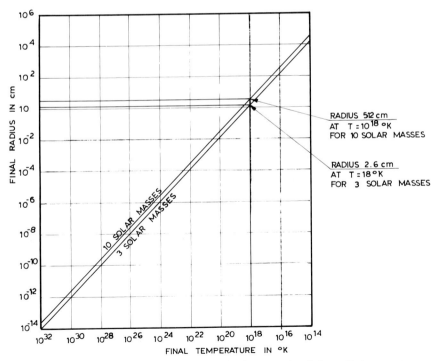

Figure 24.21. Radius of black holes of 3 and 10 solar masses in relation to final temperature.

of 520 cm and a temperature of 10^{16} K. This compares with the 18 km diameter of the Schwarzschild event horizon in the classical theory.

All theories of black holes will remain hypotheses until experimental proof is achieved. However, it will be difficult to find a flaw in my theory, which is based on known and proven physics of high-energy particles, their transmutations to radiation, as well as the known and proven equations for gravitational and thermal radiation energies.

There is no question in my mind that quasars have a spinning black hole in their center. Spinning black holes are extremely efficient machines for converting the rest mass of accreted gas into radiant energy. The intense radiation would come from an effective "photosphere" at an approximate radius of 2×10^{15} cm. Quasars are small in size. This can be determined from variations in light and radiation in 1 light-year, equal to 300 AU. This would correspond to having 100 galaxies within a volume of the solar system.

A black hole of a billion solar masses has a radius of approximately 10 AU or 1 light-hour. Infalling gas heats up to radiate X rays. It also forms an accretion disk. The radiation from the disk is responsible for the enormous quasar luminosity.

An average quasar is brighter than 300 billion stars. In a cycling universe, black holes from a previous cycle would agglomerate in the collapsed fireball (Velan theory) and after explosion, could easily be the instigators in the creation of early quasars by pulling in adjacent mass and radiation.

25

The Fate of the Universe

The future of the universe depends on whether the present expansion is slowing down and whether, eventually, the gravitational forces acting inward will overcome the kinetic energy induced in the primordial mass by the explosion of the fireball which has given the universe the push of its expansion outward. If this is the case, the expansion of the universe will eventually stop, initiating an accelerating contraction. All of this depends on the density and the total mass of the universe.

In my theory of creation with the finite, primordial fireball and its gigantic explosion caused by the overwhelming thermal forces of the hot plasma against gravitation, I assumed that there is enough mass in the universe to close it and that the density is larger than critical. Under these circumstances, the present expansion rate is gradually decreasing and the universe will eventually come to a halt. The universe will remain static for a short while and then, under the pull of its own gravitational force, it will reverse its movement and start to collapse.

25.1. THE MISSING MASS

After many years of observation, using visual, radio, infrared, and gamma ray telescopes, no more than 10% of the critical mass in the form of luminous matter required to close the universe was discovered. Cosmologists faced with these results concentrated their efforts on mysterious cold, dark particles of matter which, supposedly, do not radiate and therefore could not be detected. As over 90% of the universe appeared to be "missing," many invisible heavy particles have been invented such as axioms and supersymmetry particles which would be "mirror" images of subatomic particles, such as quarks and electrons but with much, much heavier masses. None of these cold, dark particles have been discovered and there is little solid scientific basis for their existence. Nevertheless, the behavior of galaxies and rotation of stars around the center of galaxies indicate that there must be much more mass than what we observe. Also, the expanding universe has its own escape velocity like all subjects affected by gravity.

If all matter in the universe were to move with the escape velocity, the universe would expand forever. As already mentioned, this depends on density. The larger the density is, the more effective the gravitational pull on the expanding matter and the larger the escape velocity must be. A universe containing more than the critical density will ultimately collapse, and a low-density universe will expand forever. In-between is the critical density at which the expansion balances the gravitational attraction.

The relationship between the present density of the universe d_0 and the critical density d_{crit} is expressed by the cosmological density parameter

$$\Omega_0 = \frac{d_0}{d_{\text{crit}}}$$

If $\Omega_0 > 1$, the universe is closed and will collapse. The movements of stars, mainly those that orbit the galactic center at the largest distances, indicate that the total mass of all galaxies exceeds the visible mass by at least 10–20 times. There is up to 20 times more dark mass than luminous matter in galaxies.

The dark matter is hiding in dead stars such as old white and black dwarfs, neutron stars, black holes, intergalactic dust, and vast halos of nearly all galaxies. We can calculate the total actual mass M_G of any galaxy once we know its orbital speed v and the distance R from the center. The total mass M_G in grams of a galaxy can then be calculated from the equation:

$$M_G = \frac{v^2 R}{G} \qquad (g)$$

where G is Newton's gravitational constant.

In the years 1989–1991, enormous concentrations of mass have been discovered, causing the theories of the cold, dark "missing" mass to fade. Until recently, the largest known galaxy was Markarion 348, 1.3 million light-years wide or 13 times larger than our own Milky Way and 100,000 light-years across. A newly discovered galaxy in the cluster Abell 2029 is about 6 million light-years in diameter and contains more than 100 trillion stars.

In addition, giant agglomerations of supergalaxies, at least 500 million light-years long and 15 million light-years wide, called great walls or the great attractors, have been observed and which most probably are evenly spread through the entire universe (Figure 25.1). These superconcentrations of matter are exposing to us the so far missing mass.

Thirteen more great walls and the first cluster of great walls have also been seen stretching out in a line over 7 billion light-years. This multiwall phenomenon

The Fate of the Universe

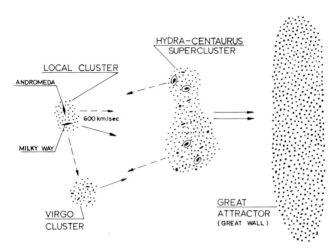

Figure 25.1. The local galaxy group, Virgo, and Hydra-Centaurus and many other clusters are gravitationally pulled by the superconcentration of clusters, the great wall.

of enormous concentration of mass has been discovered by a joint American and British team. They also found a remarkable regularity in distances between the individual walls.

To these gigantic clusters of clusters of galaxies we must add the enormous mass contained in large clusters of powerful quasars formed earlier in the universe. Their recent appearance in infrared and gamma ray telescopes at an age of about 15 billion years reinforces my own theory of creation, estimating the age of the universe to be 18 billion years. It strains and threatens, on the other hand, to break up all classical cosmological models which assign 10 or a maximum of 15 billion years to the time of the big bang explosion and the birth of the universe from the singularity.

Another excellent candidate for dark, nonluminous matter are the neutrinos, which fill the universe, interact very little with matter, and are therefore invisible. Neutrinos can pass through millions of kilometers of lead without interference. At the time when the large structures were not known or even theoretically predicted, neutrinos were not considered candidates for cold, dark matter due to the relativistic speeds at which they move. It was difficult to imagine that they would aggregate gravitationally in individual galaxies. When neutrinos decoupled at 10^{11} K, they were extremely hot and moved at a velocity close to the speed of light, not interacting with the matter and energy of space.

As the universe expanded, the neutrinos cooled and their speed considerably reduced. The slow-moving neutrinos, at 15–30,000 km/sec, bounced into matter, stopped, and joined with the baryons and electrons, considerably adding to the mass of the large agglomeration of matter, the great walls and attractors. After the discovery of the superstructures, there is a great possibility that they are, to a large extent, a part of the mass of great walls and attractors. There is an estimated equal number of neutrinos in the universe as photons, about 10^9 to 1 baryon or 10^{89} neutrinos in total.

For years, experiments have been continuing in an effort to determine whether neutrinos are massless like photons or whether they have a small mass. There are three types of neutrinos. The "lightest" and most common is the electron-neutrino (v_e); the others are the muon neutrino (v_μ) and tau neutrino (v_τ). Measurements in Los Alamos, Zurich, Munich, Moscow, and Tokyo indicate that neutrinos have an extremely small mass, if at all, of 18–30 eV. The lowest limit is 3 eV. The estimated critical mass M_{cu} in the universe used to calculate the events during the birth and evolution of the universe is

$$M_{cu} = 5.68 \times 10^{56} \text{ g}$$

If we assume that the electron-neutrino has the lowest possible weight of 2.5 eV, the total mass contribution would be

$$M_{v_e} = 2.5 \times 10^{89} \text{ eV}$$

As 1 eV = 1.782×10^{-33} g,

$$M_{v_e} = 2.5 \times 10^{89} \times 1.782 \times 10^{-33} = 4.455 \times 10^{56} \text{ g}$$

The contributions of neutrinos to the total mass of $M_{cu} = 5.68 \times 10^{56}$ g would be 79%. As can be seen, even a minute weight of 4.455×10^{-33} for 1 neutrino would represent a considerable part of the total mass.

25.2. THE CRITICAL DENSITY

It is important now to establish all of the important parameters such as the age, total mass, density, and radius of the universe, the critical density being the most essential one. We have already calculated the critical density but must repeat the exercise within the context of the fate of the universe.

The mass M_u of the universe, which is finite and which can be considered to be a sphere, can be calculated from its volume times the cosmic density d_0. Considering that R_0 is its radius

$$M_u = \frac{4\pi R_0^3}{3} d_0 \text{ g}$$

In accordance with the Newton theory of gravitation, the potential energy E_{pG} of a galaxy of mass m at the surface of the sphere with radius R_0 is

$$E_{pG} = \frac{mM_u G}{R_0} = \frac{4\pi m R_0^2 d_0 G}{3}$$

The velocity v of the galaxy can be determined from the known Hubble equation, H_0 being the Hubble constant of expansion:

$$v = H_0 R_0$$

The kinetic energy of mass m at velocity v is

$$E_k = \frac{mv^2}{2} = \tfrac{1}{2} m H_0^2 R_0^2$$

The total energy of the galaxy at the surface of the sphere representing the universe is

$$E_T = E_{pG} + E_k = \tfrac{1}{2} m H_0^2 R_0^2 - \frac{4\pi m R_0^2 d_0 G}{3}$$

$$= m R_0^2 \left(\tfrac{1}{2} H_0^2 - \tfrac{4}{3} \pi d_0 G \right)$$

For the galaxy's kinetic energy to be just in balance with gravity, the total energy E_T must be 0 or:

$$0 = mR^2 (\tfrac{1}{2} H_0^2 - \tfrac{4}{3} \pi d_0 G)$$

We divide by MR^2 and obtain

$$\tfrac{1}{2} H_0^2 = \tfrac{4}{3} \pi d_0 G$$

Therefore, for a balance between the kinetic energy and gravitation, the density d_0 has to become critical d_{crit} or, as already established,

$$d_{crit} = \frac{3 H_0^2}{8\pi G}$$

$$= \frac{3 H_0^2}{8\pi (6.67 \times 10^{-8})}$$

The Hubble constant $H_0 = 17.8$ km/sec per million light-years, 1 light-year $= 9.5 \times 10^{17}$ cm, and 1 million light-years $= 9.5 \times 10^{23}$ cm or $\sim 10^{24}$ cm:

$$d_{crit} = \frac{3}{8\pi} \left(\frac{17.8 \times 10^5}{10^{24}} \right)^2 \times \frac{1}{6.67 \times 10^{-8}} = 5.67 \times \frac{10^{10}}{10^{48} \times 10^{-8}}$$

$$= 5.67 \times 10^{-30} \text{ g/cm}^3$$

We have established that the critical density must be $d_{crit} = 5.67 \times 10^{-30}$ g/cm^3.

We must now establish the other important parameters such as the radius R_0 of the universe today and its age.

If the universe could move at the speed of light c, its radius would be

$$v = R_0 H_0$$

$$R_0 = \frac{v}{H_0} = \frac{c}{H_0}$$

With $c = 3 \times 10^{10}$ cm/sec and $H_0 = (17.8 \times 10^5$ cm/sec$)/10^{24}$ cm, the present radius of the universe would be

$$R_0 = \frac{c}{H_0} = \frac{3 \times 10^{10} \times 10^{24}}{17.8 \times 10^5} = 1.68 \times 10^{28} \text{ cm}$$

Assuming again that the universe expanded with the speed of light (which as we know is impossible since matter cannot move at this ultimate speed), the age of the universe, which is the inverse of H_0 would be:

$$\text{Age} = \frac{1}{H_0} = \frac{10^{24}}{17.8 \times 10^5}$$
$$= 5.62 \times 10^{17} \text{ sec}$$

As 1 year $= 3.15 \times 10^7$ sec,

$$\text{Age} = \frac{5.62 \times 10^{17}}{3.15 \times 10^7} = 1.78 \times 10^{10} = 17.8 \text{ billion years}$$

With the deceleration parameter of 1.5 today

$$q_0 = 1.5$$

the present recession speed of the universe was reduced from near the speed of light at the big bang to an estimated 200,000 km/sec. The estimated speed at the explosion was $0.999c$ or 299,000 km/sec.

The average speed v over the 18 billion years was

$$v = \frac{3 \times 10^{10} + 2 \times 10^{10}}{2} = 2.5 \times 10^{10} \text{ cm/sec (250,000 km/sec)}$$

The present radius R_0 of the universe should therefore be

$$R_0 = \frac{v}{H_0} = \frac{2.5 \times 10^{10} \times 10^{24}}{17.8 \times 10^5} = 0.14 \times 10^{29} = 1.4 \times 10^{28} \text{ cm}$$

As 1 light-year $= 0.946 \times 10^{18}$ cm

$$R_0 = 14.8 \times 10^9 \text{ light-years}$$
$$= 1.4 \times 10^{28} \text{ cm}$$

The Fate of the Universe

The density today can be calculated from:

$$d_0 = \frac{3}{8\pi R_0^2} + \frac{3H_0^2}{38}$$

$$d_0 = 6 \times 10^{-56} \text{ cm}^{-2} + \tfrac{3}{8}\left(\frac{17.8 \times 10^5}{10^{24}}\right)^2$$

$$= 6 \times 10^{-56} \text{ cm}^{-2} + 37 \times 10^{-38} \text{ sec}^{-2}$$

$$= 6 \times 10^{-56} \text{ cm}^{-2} + 3.7 \times 10^{-37} \text{ sec}^2$$

$$d_0 = (6 + 4.07) \times 10^{-56} \text{ cm}^{-2}$$

$$= 10.07 \times 10^{-56} \text{ cm}^{-2}$$

$$= 13.54 \times 10^{-30} \text{ g/cm}^3$$

The present density is $d_0 = 13.54 \times 10^{-30}$ g/cm³ and the cosmological density parameter Ω_0 is the relationship between the present density and the critical density. It must be larger than 1 for the universe to slow down its expansion and come to a halt:

$$\Omega_0 = \frac{d_0}{d_{\text{crit}}} = \frac{13.54 \times 10^{-30}}{5.67 \times 10^{-30}} = 2.39 \quad \text{or} \quad > 1$$

We must now determine the remaining data in order to assess the fate of the universe such as the radius at maximum expansion R_{max}, its density, the time from start to maximum and from start to full recontraction.

R_{max} is a sum of the present radius R_0 plus the final expansion R_{FE} from now to a total halt or $R_{\text{max}} = R_0 + R_{\text{FE}}$. We can calculate R_{FE} from

$$R_{\text{FE}} = \frac{V_{\text{FE}}}{H_{\text{FE}}}$$

If the present expansion velocity is 2×10^{10} cm/sec and the universe will come to a halt at 0 velocity, it will expand at an average velocity of $(200 + 0)/2 = 100$ cm/sec.

The present Hubble constant of 17.8 km/sec per million light-years will become $H_{\text{FE}} = 0$. The average value, therefore, will be: $H_{\text{FE}} = 8.0$ km/sec per million light-years. We can therefore calculate the final expansion distance from

$$R_{\text{FE}} = \frac{V_{\text{FE}}}{H_{\text{FE}}} = \frac{10^{10} \times 10^{24}}{8.9 \times 10^5} = 1.12 \times 10^{28} \text{ cm}$$

The maximum expansion radius will therefore be

$$R_{\text{max}} = R_0 + R_{\text{FE}} = 1.4 \times 10^{28} + 1.12 \times 10^{28}$$

$$= 2.52 \times 10^{28} \text{ cm}$$

$$= 26.6 \times 10^9 \text{ light-years}$$

The age A_{ME} of the universe at maximum expansion can be calculated from the average Hubble constant H_{AE}:

$$A_{ME} = \frac{1}{H_{AE}}$$

As H_{AE} = 8.9 km/sec per million light-years and 1 year = 3.15 × 10⁷ sec,

$$A_{ME} = \frac{10^{24}}{8.9 \times 10^5} = 0.112 \times 10^{19} \text{ sec}$$

$$= \frac{0.112 \times 10^{19}}{3.15 \times 10^7} = 0.0355 \times 10^{12} \text{ years}$$

$$= 35.5 \times 10^9 \text{ years}$$

It will therefore take 17.5 billion more years for the universe to come to a full stop and then recollapse (present age: 18 billion years). The whole cycle from the explosion of the fireball to final recontraction into a fireball will take 71 billion years (Figure 25.2).

The density at maximum expansion d_{max} is

$$d_{max} = \frac{3}{8\pi R_{max}^2} + \frac{3H_0^2}{8}$$

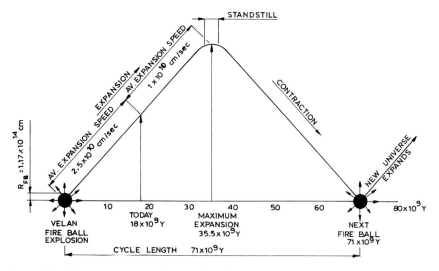

Figure 25.2. Schematic expansion and contraction of the universe on the basis of the Velan cosmological model.

As H_0 becomes 0

$$d_{max} = \frac{3}{8\pi R_{max}^2} = \frac{0.119}{(2.52 \times 10^{28})^2}$$
$$= 0.0187 \times 10^{-56} = 1.87 \times 10^{-58}$$
$$= 2.52 \times 10^{-30} \text{ g/cm}^3$$

Table 25.1 lists all of the important data compatible with the Velan cosmological model of a closed universe, complying with the Einstein cosmological concept but eliminating the singularity. I must emphasize that all data are to be considered approximate. The data are based on the best possible analysis of all observational experience combined with logical consequences of the Velan cosmological model for a closed universe, created from the primordial radiation energy of the cosmos. The main purpose of the numbers is to serve as an illustrative background for the theory.

According to this model, the universe will continue to expand though with continuously reduced velocity over a period of 17.5 billion years until it comes to a complete halt. During this period, the background radiation in the form of photons and neutrinos will continue expanding and cool down in inverse proportion to the size of the universe until its temperature drops from the present 3 K to 1.6 K and its energy, once dominating the universe, will be minimal. As the universe, in accordance with our calculation, will be 1.8 times larger, the temperature of the background radiation will drop to 1.6 K above absolute zero. Many stars and even complete galaxies will exhaust their sources of energy and will become dead, cold masses of matter and radiation. Many galaxies will be absorbed into the gigantic concentration of mass in the form of the great walls and great attractors.

Table 25.1. Important Data concerning the Universe as Depicted in the Velan Cosmological Model

Radius at explosion of the fireball	$R_{FB} = 1.17 \times 10^{14}$ cm
Radius today	$R_0 = 1.4 \times 10^{28}$ cm
Radius at maximum expansion	$R_{ME} = 2.52 \times 10^{28}$ cm
Density at explosion	$d_{FB} = 1.3 \times 10^{15}$ g/cm^3
Density today	$d_0 = 13.54 \times 10^{-30}$ g/cm^3
Density at maximum	$d_{max} = 2.52 \times 10^{-30}$ g/cm^3
Time from explosion to today	$t_0 = 18 \times 10^9$ years
Time from start to maximum expansion	$t_{ME} = 35.5 \times 10^9$ years
Time from start to final collapse	$t_{FC} = 71 \times 10^9$ years
Amount of matter	$M_u = 5.68 \times 10^{56}$ g
Cosmological density parameter $\Omega_0 = d_0/d_{crit}$	$\Omega_0 = 2.39 > 1$

Trillions of more dwarfs, neutron stars, and black holes will be added to the inventory of the expanding universe when slowly the expansion will stop and everything will come to a short pause.

As discussed in detail in an earlier chapter, our sun and planetary system will turn into cinders long before the universe comes to a halt. In fact, civilization and all life on earth will be extinguished within a short period of time, speaking in cosmic terms, of 500 million years during the first substantial expansion of the sun. Five billion years later, undergoing a supergiant extension, the sun itself will become a small white dwarf. The planet earth, as such, may survive and orbit the white dwarf at a much larger orbit—a lonely, cold, dead, and invisible planet. The white dwarf will itself ultimately become a dead, cold, black, dwarf star, joining trillions and trillions of similar bodies in the expanding and aging universe, which by now is slowing down its expansion pace.

There is a theoretical possibility that somewhere in the distant galaxies, younger in age and still containing stars with thermonuclear reaction, another intelligent civilization, even more advanced than the humans just extinguished, is inhabiting a safe planetary system. If so, they may witness the decelerating expansion of the universe and perhaps even the early stages of retraction. By this time after more than 35 billion years, it is possible that all thermonuclear reactions in stars will come to an end and the galaxies will continue only as giant dead bodies of nonluminous mass and radiation.

If my calculations supported by the discovery of quasars 15 billion years old are correct, the universe today is 18 billion years old. It will expand for 17.5 billion years more, come to a halt for a little while longer, and then begin to collapse.

The gravitational energy of the universe with mass $M_u = 5.68 \times 10^{56}$ g, when coming to a complete stop at maximum expansion, can be calculated from

$$E_{\text{GRAV}} = \frac{GM_u^2}{R_{\text{ME}}} \quad \text{(erg)} \tag{18.1}$$

where the radius of the universe at maximum expansion is

$$R_{\text{ME}} = 2.52 \times 10^{28} \text{ cm}$$

and G, the gravitational constant, $= 6.673 \times 10^{-8}$ cm^3/g sec^2:

$$E_{\text{GRAV}} = \frac{6.673 \times 10^{-8} \times (5.68 \times 10^{56})^2}{2.52 \times 10^{28}} = 8.5 \times 10^{77} \text{ erg}$$

As the kinetic energy of the universe coming to a full stop drops to 0 ($M_u v^2/2$, $v = 0$) and the temperature drops to a low of 1.6 K from 3 K today, there will be no kinetic or thermal forces to oppose the gravitational collapse which will follow. Gravitation will become a dominant, unopposed force in the universe for the next 20–24 billion years.

The contraction cycle which will start will be a reverse in time of the previous expansion period. After about 17.5 billion years of reverse movement, the universe will retract to an approximately similar size as today, with a radius of $R_0 = 1.4 \times 10^{28}$ cm, density of $d_0 = 13.54 \times 10^{-30}$ g/cm^3, and a temperature of 3 K.

After a long period of about 18 billion years, it would be approaching the state of a fireball, very similar in content and structure to the primordial fireball with a radius of $R_{FB} = 1.17 \times 10^{14}$ cm, a quark–electron density of $d_{FB} = 1.3 \times 10^{15}$ g/cm^3, and a temperature of 10^{26} K. Soon after, a big bang explosion will set up the universe on a new cycle of expansion and glorious creativity.

Galaxies and stars will be created and finally an even more intelligent species of humans may inhabit a lonely planet and try to unravel the secret of the universe, all starting anew. Life and death seem to be an irreversible cycle in the cosmos.

25.3. LANDMARK EVENTS IN THE HISTORY OF THE RECOLLAPSING UNIVERSE

All events which will follow depend on the temperature of the background radiation and matter, which heats up in an inverse relationship to the size of the universe or its declining radius R, $t \sim 1/R$ (Figure 25.3).

$T = 3 K$

After 17.5 billion years when the universe has contracted to 0.555 of the maximum expansion size, its radius will be close to today's $R_0 = 1.4 \times 10^{28}$ cm and the temperature will have heated up from 1.6 K to 3 K.

$T = 300 K$

When the universe contracts to $R = 1.4 \times 10^{26}$ cm, the temperature will rise 100-fold to 300 K. The radiation energy will still be weak and matter will dominate the collapsing universe. Distances between galaxies and stars will remain substantial and there will be few collisions or encounters.

$T = 3 \times 10^3 K$

The first major event will take place when the universe has contracted $R = 1.4 \times 10^{25}$ cm and the temperature has risen to 3000 K. The sky will become intensively bright, close to the blinding light coming from our sun today, even during nighttime.

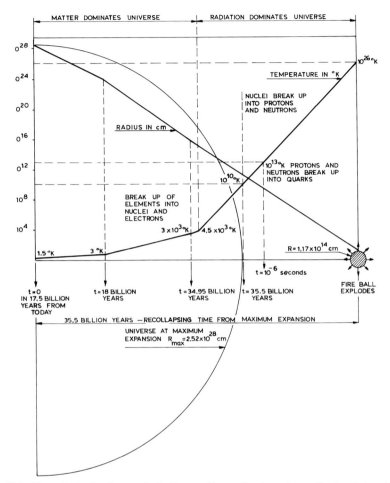

Figure 25.3. The major landmark events in the history of the recollapsing universe. Total cycle from birth to recollapse is 71 billion years.

All molecules in stars, planets, and interstellar space will start to break up into free electrons and atomic nuclei. All elements from hydrogen and helium to heavy elements will cease to exist. The universe will change into a unified mass of nuclei, electrons, neutrinos, and radiation. This will take place approximately 34.95 billion years from the beginning of contraction.

From Eqs. (20.1), (20.7), (20.8), and (20.10), we can calculate the density

The Fate of the Universe

and time from full collapse into the fireball, as well as the energy of radiation or photons.

The energy of particles of matter equal to the rest mass of particles remains constant. The thermal energy of radiation increases with temperature to the fourth power (T^4). The density is

$$d = 0.84 \times 10^{-35} \times T^4 \quad (\text{erg/cm}^3) \tag{20.1}$$

The time during matter-dominated universe from collapse is

$$t_m = 0.089 \left(\frac{10^8}{d}\right)^{1/2} \quad (\text{sec}) \tag{20.8}$$

The time during radiation-dominated universe is

$$t_r = 0.067 \left(\frac{10^8}{dc}\right)^{1/2} \quad (\text{sec}) \tag{20.7}$$

The energy of 1 photon is

$$E_{1ph} = 0.232 \times 10^{-3} \times T \quad (\text{eV})$$

The total energy of 10^9 photons compared to the energy of 1 proton is

$$E_{phT} = 0.232 \times 10^6 \times T \quad (\text{eV})$$

At this temperature, the energy of radiation will still be smaller than that of matter which dominated the universe:

$$E_{1ph} = 0.232 \times 10^{-3} \times 3 \times 10^3 = 0.69 \text{ eV}$$
$$E_{phT} = 0.69 \times 10^9 = 690 \times 10^6 \text{ eV} \quad < \quad 933 \times 10^6 \text{ eV}$$

$T = 4.5 \times 10^3 \, K$

When the universe will heat up to 4500 K due to contraction, reducing its size to $R = 0.934 \times 10^{25}$ cm, radiation energy will become more powerful than the energy of matter. At 4500 K, the energy of radiation will be

$$E_{1ph} = 0.232 \times 10^{-3} \times 4.5 \times 10^3 = 1.044 \text{ eV}$$
$$E_{phT} = 1.044 \times 10^9 \text{ eV} = 1044 \times 10^6 \text{ eV} \quad > \quad E_{PR} = 933 \times 10^6 \text{ eV}$$

Overall, radiation will become more powerful than matter.

The universe, now in full thermal equilibrium, will become dominated by radiation. The gravitational energy is steadily increasing due to the diminishing size and radius of the universe. The kinetic energy is in turn transferred into more intensive collisions of particles and increase in thermal and pressure forces which will soon start opposing the gravitational collapse.

$T = 10^{10} K$

After another 543,000 years, the temperature will rise to 10^{10} K. All atomic nuclei will break up into protons and neutrons, destroying all of the work of nucleosynthesis and thermonuclear activities in the centers of stars.

Entire galaxies and their trillions upon trillions of stars, dead bodies, in the form of dwarfs, neutron stars, and intergalactic matter will all combine again into a single, extremely hot cosmic soup of protons, neutrons, electrons, neutrinos, and radiation.

Black holes will be part of the collapsing universe but will remain intact. In some cases they will absorb some radiation and matter and cause local discontinuities of density and rapid reduction of pressure in their vicinity.

Some of the black holes may be absorbed by others, increasing their horizon. It is difficult to predict the exact behavior of black holes during the collapsing cycle. They will certainly play a major role, however, when the collapsed universe explodes and commences its next expansion.

It is possible that the presence of black holes will cause an early formation of primordial quasars, attracting a large quantity of matter falling into a black hole in the center of the forming quasars. The outward-acting thermal forces and radiation pressure will substantially increase in this cosmic soup and oppose the inward-acting forces of gravitation.

The radius of the universe at 10^{10} K would be substantially smaller:

$$\frac{10^{10}}{4.5 \times 10^3} = 2.3 \times 10^6$$

$$R = \frac{0.934 \times 10^{25}}{2.3 \times 10^6} = 4 \times 10^{18} \text{ cm}$$

The gravitational energy will be a billion times stronger than at the time of maximum expansion of the cold universe at 1.6 K and $R = 2.52 \times 10^{28}$ cm.

$$E_{\text{GRAV}} = \frac{GM_u^2}{R_{10^{10}}} = \frac{6.673 \times 10^{-8} (5.68 \times 10^{56})^2}{4 \times 10^{18}} = 5.4 \times 10^{87} \text{ erg}$$

$$\frac{5.4 \times 10^{87} (\text{at } 10^{10} \text{K})}{8.5 \times 10^{77} \text{ (at maximum expansion)}} = 6.3 \times 10^9$$

The thermal force of the dominating radiation will also substantially increase, opposing the gravitation.

At 10^{10} K the thermal energy of electromagnetic radiation can be calculated from

$$\begin{aligned} E_{\text{EMR}} &= aT^4 \quad (\text{erg/cm}^3) \\ &= 7.564 \times 10^{-15} \times (10^{10})^4 \\ &= 7.564 \times 10^{25} \text{ erg/cm}^3 \end{aligned}$$

For total energy, we must multiply E_{EMR} by the volume of photons. The number of photons in 1 cm³ is

$$N = 20.3 \times T^3 = 20.3 \times (10^{10})^3$$
$$= 10.3 \times 10^{30} \text{ photons in 1 cm}^3$$

The number of photons in the universe is 10^{89}:

$$\text{Volume } F_{EMR} = \frac{10^{89}}{20.3 \times 10^{30}} = 5 \times 10^{57} \text{ cm}^3$$

The total energy of radiation is

$$E_{EMR/T} = 7.564 \times 10^{25} \times 5 \times 10^{57} = 37 \times 10^{82} \text{ erg}$$
$$= 3.7 \times 10^{83} < E_{GRAV} \; 5.4 \times 10^{87}$$

Gravitation is still more than 15,000 times stronger than the thermal forces. The collapse will continue.

One year before the final collapse of the universe into a fireball which will later explode, all galaxies will overlap and all stars will break up into interstellar matter which, in reality, will form an extremely hot soup of quarks, electrons, neutrinos, and radiation. All of the creative work of billions of years will be undone. The thermal energy of radiation and its pressure will be starting to effectively counteract the gravitational forces.

$T = 10^{13} K$

An important event will take place when the universe contracts into a fireball with a radius of $R = 4 \times 10^{15}$ cm and temperature of 10^{13} K. The strong force keeping the quarks confined inside of protons and neutrons will uncouple and quarks, the constituents of protons and neutrons, will become free particles. The extremely hot cosmic soup will now contain quarks, electrons, and extremely energetic neutrinos and photons dominating the universe. This will take place only 10^{-6} sec before the fireball explodes again and puts the universe on another cycle of expansion and brilliant future of progress and development.

$T = 10^{26} K$

In less than a millionth of a second, the fireball having a radius of about 1.17 $\times 10^{14}$ cm will react to the incredibly high temperature of 10^{26} K and the radiation dominating the universe will achieve an energy level of 10^{18} GeV. The density of the fireball will achieve the quark–electron density of 1.3×10^{15} g/cm³ and another big bang explosion will follow (Figures 19.1 and 19.2).

By the time the entire fireball reaches the quark–electron density, the core will be compressed far beyond this density level. All of the electrons, quarks, and other particles in the fireball will merge to form a sort of single gigantic nucleus. A spoonful of such matter has approximately the same mass as all of the buildings in Montreal combined. In this form, particles of matter show a powerful resistance to further compression. This, however, does not stop the outside layers of particles in the fireball from imploding further and exerting more squeezing power on the core.

At the surface of the hard core, the particles stop suddenly but not fully. The compressibility of elementary particles is low at nuclear density but not zero. The momentum of the infalling particles (mass m × velocity v), being close to the speed of light, will compress the central sphere to perhaps a density 4–5 times that of equilibrium, which we can call the point of the "maximum squeeze." By that time the thermal energy of particles and radiation, as calculated, will reach a level higher than the total gravitational energy and will also exert considerable pressure against the gravitational forces.

While the internal pressure in the center of the sun's core is 10 billion kg/cm^2, the pressure in the interior of the fireball will be about 10^{31} kg/cm^2 or 10 million, trillion, trillion kg/cm^2. The core, after the "maximum squeeze," will bounce back like a rubber ball that was compressed. The bounce will set off enormous shock waves which, together with the overpowering internal forces, will create mainly the energy of the electromagnetic radiation and will result in a titanic cosmic explosion, as shown schematically in Figures 19.1 and 19.2.

The thermal forces of radiation calculated on page 223 will overpower the gravitational energy at the high temperature of the highly compressed fireball:

$$E_{EMR/T} = 6.443 \times 10^{100} \text{ erg} > E_{GRAV} = 287 \times 10^{90} \text{ erg}$$

The universe will start a new cycle of expansion, similar to the previous bounce. Galaxies will be formed; later, stars and planetary systems and perhaps intelligent life. We may be living presently in a universe that collapsed and rebounded. Nevertheless, even with this cosmological model of repeated expansions, retractions, and rebounce, the universe must have once been created for the first time.

It is this process—the creation of the universe from primordial radiation—that was the purpose and aim of this book.

Epilogue

While this book was in the process of being published, the world was confronted in April 1992 with startling press headlines concerning recent important cosmological observations made by two NASA orbiters. There were such headlines as

- "Looking at God"
- "A satellite finds the largest and oldest structures ever observed"
- "Echoes of the big bang"
- "Evidence of how the universe took shape 15 billion years ago"
- "The big bang–singularity theory now proven"
- "How the missing link of creation was discovered"
- "Missing link between big bang and matter"
- "Proof positive for the big bang believers"
- "Astronomers find Holy Grail of the cosmos in first sign of creation"
- "Clouds of matter found by satellite"

As these discoveries have an influence on my theory of creation described in this book, I would like to put the findings in proper perspective for the readers. None of the headlines accurately reflects the true meaning of the discoveries.

Cosmic Background Microwaves and the NASA Orbiter COBE

The NASA Cosmic Background Radiation Explorer, COBE, launched in November 1989, confirmed first that the hot radiation of the early universe, which decoupled from matter approximately 500,000 years after the big bang explosion and first discovered in 1964 by Penzias and Wilson, expanded and cooled down with the universe to become a very weak microwave radiation, about 100 million times fainter than the heat produced by a candle.

COBE also confirmed first that the microwave radiation's temperature is 2.735 K or $-270\,°C$, that this temperature is uniform across the entire sky, and, therefore, left the early universe or fireball which had uniform temperature and density. Cosmologists, believing in the classical theory of the big bang of a singularity, had trouble reconciling this picture of early smoothness with today's universe of billions of galaxies and stars. Until the time of decoupling, the universe was opaque, as radiation in the hot fireball could not move and was scattered by matter. Therefore, everything before the release of radiation cannot be observed and must be inferred. To ask about the origin of the singularity or what was before, remains a meaningless question as there was no before. It is an inpenetrable mystery, beyond human reach. There was no time, no space, only nothingness. Matter, space-time, and the universe started with the explosion of the singularity for no apparent reason. The mystery of the singularity remains. It was, seemingly, a mathematical point of zero space containing all matter and radiation of the universe, compressed to infinite density and temperature.

Until early April 1992, the microwave glow seemed to be evenly spread across the sky. How then did the universe become clumpier with local differences in density which caused gravity to accumulate large masses and collapse them into protogalaxies and, later, galaxies and stars?

According to quantum mechanics and the Heisenberg uncertainty principle, well documented and proven, nothing is perfectly smooth; so a slight unevenness must have already existed in the small fireball—a universe—even in those early stages. It is this unevenness in the temperature of the microwave background radiation which the COBE orbiter subsequently found, in hundreds of thousands of recorded measurements, and which Dr. Smoot believes he has mapped. *Nothing more, nothing less.* He did not "discover or find the largest and oldest structures ever observed—evidence of how the universe took shape 15 billion years ago," as reported in the May 4, 1992 issue of *Time* magazine.

In addition, the anomalies recorded by COBE so far of 30 millionths or 0.000030 of a degree cannot be the seeds of clusters and galaxies observed today, as COBE observes large areas of the sky, $7°$ or more. You can, however, assume that if there were fluctuations in density on the scale that COBE can "see," there would be similar fluctuations on smaller scales which could eventually explain the formation of galaxies. These observations, however, must still be made.

There is another problem as well. The features "observed" by COBE are wide, but not deep. The high recordings are only a thousandth of a percent higher than the lows and therefore the density differences are very small. Although the results are remarkable, the promoters of the classical theory of creation must turn to the so-called "cold, invisible dark matter" of large mysterious dark particles in order to amplify the gravitational effects of visible seeds by the presence of something which does not interact with normal matter, as we know it, in order to explain the future gravitational clumping of matter into galaxies.

TIME	THE CLASSICAL THEORY OF THE BIG-BANG OF A SINGULARITY	THE VELAN COMPLETE THEORY OF CREATION "NO SINGULARITY"
BEFORE THE SINGULARITY	*NOTHINGNESS* A meaningless question as there was **no before**. It is an impenetrable mystery, beyond human reach. There was no time, no space; only nothingness	An infinite multi-universe cosmos containing space and universes. In the inter-universe space flows a high-energy, primordial radiation carried by powerful γ rays, at speeds greater than the speed of light. It carries the seeds of creation. Each universe follows the laws of the cosmos from its birth, its expansion through its own horizon determined by its total mass, its evolution, then gravitational implosion and finally explosion into a new cycle of expansion and development. The multi-universe cosmos in its vastness and may be infinity is the witness to the glory of the Creator.
SINGULARITY APPEARS	*SINGULARITY* ● **NO TIME - NO SPACE YET** A mathematical point of 0-space with infinite density of ? some say of pure energy, others, containing all the matter and radiation of the universe compressed to infinite density at infinite temperature. No explanation of how it was created. All physical laws, as we know them, including Einstein's equations, break down.	

PEERING BACK TO GENESIS

TIME	THE CLASSICAL THEORY OF THE BIG-BANG OF A SINGULARITY	THE VELAN COMPLETE THEORY OF CREATION "NO SINGULARITY"
15 BILLION YEARS AGO	 ### TIME 0 – NO SPACE The singularity containing matter and radiation of infinite density explodes for no apparent reason. This is the beginning of matter, time, space and the universe - an eccentric fantasy with little scientific justification. No laws of nature as we know them govern this event.	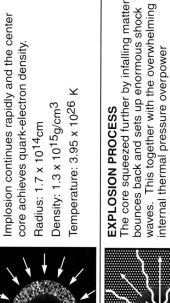 ### THE CREATION MOMENT - Principle only Violent fluctuations of superspace, in tiny cells of 10^{-33} cm, together with massive appearance of virtual particles interact with the primordial radiation which provides the rest mass for the virtual particles. Matter and photons are created.  ### GRAVITATIONAL IMPLOSION Gravitational collapse of the newly-created hot soup of quarks, electrons, neutrinos and photons occurs. Part of the primordial radiation is trapped in the fireball which explains the large quantity of photons (10^9 to 1 proton) and the presence of powerful γ rays of cosmological origin. 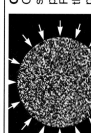 Implosion continues rapidly and the center core achieves quark-electron density. Radius: 1.7×10^{14} cm Density: 1.3×10^{15} g/cm^3 Temperature: 3.95×10^{26} K 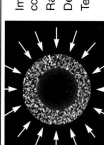 ### EXPLOSION PROCESS The core squeezed further by infalling matter bounces back and sets up enormous shock waves. This together with the overwhelming internal thermal pressure overpower gravitation and cause the explosion. The core becomes a black hole and the
18 BILLION YEARS AGO		

TIME	NEW DISCOVERIES IN 1992 BY NASA'S MICROWAVE COBE AND γ RAY ORBITERS	WHAT IT MEANS FOR THE THEORY OF THE BIG BANG OF SINGULARITY	WHAT IT MEANS FOR THE VELAN COMPLETE THEORY OF CREATION
500,000 YEARS AFTER THE BIG BANG EXPLOSION	**1. Microwave Background Radiation** When the universe started to cool down, the overwhelming radiation was becoming less powerful and quarks, the basic particles combined into protons and neutrons, formed together nuclei of atoms and still later electrons scattered by radiation combined with nuclei to form atoms of hydrogen and helium. At this point at a temperature of 3000 – 4000 K, approximately 500,000 years into expansion, the radiation decoupled from matter. The opaque universe became suddenly transparent and matter dominated. It is this escaping radiation, first discovered by Penzias and Wilson in 1964, which later expanded with the universe and cooled down to 2,735 K or -270°C, which was observed by the COBE Microwave Background radiation orbiter between 1989 and 1992. COBE discovered small temperature fluctuations of only 30 millionth (.000030) of a degree indicating differences in density of matter. **2. GAMMA (γ) RAYS** A NASA orbiter detected early in 1992 violent bursts of gamma rays, isotropic or uniform over the sky suggesting a cosmological origin.	**1a.** Although the substantial discovery is hailed as historic, the small fluctuations in temperature are not sufficiently large to explain the clumping, much later, of matter and radiation into galaxies. **1b.** Little contribution to the mystery of the singularity and its explosion or to the origin of matter and radiation. **1c.** It is not proof of the theory of the big bang of a singularity, as claimed. **2.** No explanation exists for the origin of this powerful γ ray radiation at this moment and all the theorists seem to be very confused. There exists no known process in the universe to account for such high energy γ rays even in the cataclysms that exist around quasars, pulsars, black holes, or exploding galaxies.	**1a.** It confirms that the Velan fireball had small density variations. As predicted in the theory, the decoupling of radiation (1 billion or 10^9 photons to one proton) at approximately 3000 K, resulting in a sudden and large pressure drop in the Velan fireball caused the implosion and collapse of large areas, creating shock waves and compression of gas. These were the seeds for the creation of galaxies much later. **2.** The existence of powerful γ rays were predicted in the Velan theory already in 1985. They are remnants of the primordial radiation trapped during the creation process of matter and electromagnetic radiation contained in our universe. The observed bursts of the gamma rays may be attributed to scattering by large masses of matter in the early universe preventing the escape of this powerful radiation. **The two discoveries in 1992 seem to confirm the Velan theory of creation.**

Also, only 12 years ago, scientists estimated the age of the universe to be 10 billion years. Discoveries of ancient galaxies and quasars in the last few years at distances of 12, 14, and even 15 billion light-years, gradually forced the extension of the estimated life of the universe to 15 billion years. This will have to be further extended soon, as it took at least one to two billion years for those old galaxies to form and even 3 billion years for quasars.

Now, let us again discuss decoupling of radiation. In the classical theory of the big bang, space is created by the expansion of the universe and therefore radiation, which decoupled from matter 500,000 years after the big bang, could not have escaped into "space" but must have remained at the outer region of the expanding universe.

The decoupling of radiation in the Velan fireball was more dramatic and pronounced, as space existed beyond the universe and resulted in sudden and large pressure drops in the fireball's matter. This caused the implosion and collapse of large areas, creating shock waves and compression of the hydrogen and helium gas. These were the real seeds which, much later, developed the clumping of matter into large clouds, protogalaxies, and finally into galaxies and stars. The age of the universe in my theory is estimated to be 18 billion years.

Bursts of High Energy Gamma Rays by NASA's Gamma Ray Orbiter

A major scientific discovery was made by another NASA orbiter early in 1992, which observed the universe in the range of the most powerful electromagnetic radiation carried by γ rays.

At first, in a similar way as the microwave radiation predicted by Gamov and discovered by Penzias and Wilson in 1964 came from the sky, the γ-ray bursts came from one area of the sky. Shortly after, 70 more bursts were recorded isotropically or uniformly over the sky, suggesting a cosmological origin. The observations cannot be explained by any existing theory, process, or object in the universe. The bursts, 100,000 times as luminous as the sun but invisible to the eye because of their extremely minute wavelengths, cannot be explained even by cataclysms that erupt around massive quasars, pulsars, black holes, or exploding galaxies. Specialists, after two meetings in January and April 1992, still seem to be confused. Gerald Fishman, a NASA scientist and leader of the team of the Gamma Ray Observatory declared, "You may see theories that this is a fundamentally new aspect of the universe as a whole or some new particle ray physics."

The presence of such powerful radiation in the universe, however, was predicted in 1985 by my cosmological theory. The γ rays may be remnants of the trapped primordial radiation during the creation process of the universe. One of the basic parameters of my theory is this powerful primordial radiation, carried by high-energy γ rays or photons, which flows in the interuniverse space of the cosmos and contains the "seeds" or "chromosones" of the creation of matter.

As explained in detail in this book, on rare occasions the radiation interacts with the vacuum of space and its virtual particles, creating a new universe. Part of the primordial radiation, which provided the rest mass to the created particles of matter, gets trapped in the newly created fireball by powerful gravitational forces. Professor Sir Martin Rees, director of the Institute of Astronomy at Cambridge wrote to me on May 30, 1985 after reviewing my theory of creation in its initial form: ". . . I have found this most interesting and stimulating. However, it is still going to take years of expensive and high quality development work to transform your current tentative ideas, even with the enhancement provided by your original input, into a robust and convincing overall theory."

The recorded bursts of the γ rays, which last from a fraction of a second to hundreds of seconds, are difficult to explain. However, they may be attributed to scattering by large masses of matter in the early universe, preventing the escape of the powerful radiation from the fireball.

It would appear to me that the two discoveries by NASA orbiters in 1992, if anything, serve to confirm the validity of my cosmological theory of creation.

Glossary

absolute luminosity: Total energy emitted per second by an astronomical object.

alpha particle: The nucleus of a helium atom, consisting of two protons and two neutrons. It is radiated from radioactive substances.

angstrom (Å): 10^{-8} cm; one hundred-millionth of a centimeter. Light waves have a wavelength of a few thousand angstroms.

antiparticle: Also called antimatter. Each particle has its antiparticle with opposite electrical charge, baryon and lepton number, such as electron and positron, proton and antiproton. Nature has preference for matter.

apparent luminosity: Energy actually received per second and per square centimeter from an astronomical body or flux.

asymptotic freedom: At energies above 15 GeV or distances smaller than 10^{-13} cm, quarks are free particles; under all other conditions, they are confined to baryons or mesons.

baryons: Strongly interacting particles such as protons, neutrons, and other hadrons which are unstable.

beta decay: The decay of a neutron into a proton, electron, and electron-antineutrino.

big bang: The classical cosmological theory of the birth of the universe from a singularity or mathematical point of infinite density, energy, and temperature. In the Velan theory, the primordial fireball has finite dimensions.

binary system: A system of two stars orbiting around a common center.

blackbody radiation: Radiation from a body in thermal equilibrium.

black hole: A singularity of a collapsed, dead star's core larger than 2.5 solar masses.

blueshift: A shift toward shorter wavelengths caussed by the Doppler effect of an approaching body.

Boltzmann's constant (k): A constant relating the temperature level to units of constant energy. $k = 1.3806 \times 10^{-16}$ erg/K or 0.00008617 electron volts/K.

bosons: All particles that have integral spin such as photons (spin 1), $W^{\pm}$ bosons (spin 1), and gluons—particles that mediate the four forces of nature.

charm (C): A quantum number of charm (c) quarks in a particle less anticharm quarks.

charmonium: A system containing a charm quark and an anticharm quark.

chromodynamics (QCD): The theory of strong interaction between quarks via gluons (bosons).

closed universe: A cosmological model in which space is positively curved, is finite, and may finally recollapse.

Compton wavelength: A particle in quantum mechanics has a wavelength (h/mc).

conservation law: States that the total value of a given quantity does not change in any reaction.

conservation of energy: The total energy of a system remains constant irrespective of whatever internal changes may take place.

cosmology: The study of the birth and evolution of the universe.

critical density: The minimum mass density of the universe that will cause the expansion to stop and start to contract in a reverse action.

decoupling era: The era of the universe after creation when temperature cooled down to only 3000 K. Radiation, which dominated, escaped and matter dominated the universe.

degenerate matter: Particles of matter such as electrons or neutrons, compressed to nuclear density, resist further compression by gravity. The gas at this density is called degenerate matter.

density: The amount of any quantity per cubic centimeter.

deuteron: The nucleus of heavy hydrogen, consisting of one proton and one neutron.

Doppler effect: The change in frequency of waves (as sound, light, or radio waves), caused by relative motion of the source and receiver.

electron (e): The lightest elementary particle of matter. It has a negative electrical charge. Chemical properties depend on electron interaction between molecules and atomic nuclei.

electron volt: A unit of electric energy acquired by an electron passing in an electric field of 1 volt (V). 1 eV = 1.6×10^{-9} watt/second. MeV = 10^6 eV, GeV = 10^9 eV.

elliptical galaxy: A galaxy without arms. It contains 10^7–10^{12} solar masses and is ellipsoidal in shape.

energy density: The amount of energy (in erg/cm^3) of radiation at temperature T. It equals aT^4 where a is the radiation constant $a = 7.56 \times 10^{-15}$ erg/cm^3 K^4.

erg: The unit of energy in the centimeter–gram–second system.

Fermi constant: A unit of the strength of the weak interaction. Denoted $G_F = 294$ GeV^{-2}.

fermion: A term for a particle whose spin quantum number is an odd multiple of $\frac{1}{2}$.

fine structure constant (α): A unit of electrodynamics $\alpha = e^2/hc^2 = 1/137$, where e is the charge of an electron, h is Planck's constant, and c is the speed of light.

flavor: A property that distinguishes different types of quarks (u, d, c, s, b, t) and different kinds of leptons.

frequency: The number of complete oscillations per second of an electromagnetic wave. It is measured in cycles per second (hertz).

Freidmann model: The mathematical model of the universe based on general relativity and the classical big bang theory.

galactic nucleus: Region of enhanced central stellar density; sometimes a massive black hole.

galaxy: A large mass of stars and interstellar matter, containing an average of 10^{11} solar masses. Their shape is classified as elliptical, spiral, barred spiral, or irregular.

gamma rays (γ): Photons of high energy and high frequency.

general relativity: Formulated by Einstein. Its basic idea is that gravitation is an effect of the curvature of the space-time continuum.

gluons: Massless and electrically neutral particles with spin 1. Mediate the strong interaction between quarks in nucleons and mesons as well as between nucleons in the nucleus of atoms.

gravitational lens: A large galaxy or cluster that bends light from an object in the back, causing multiple images.

gravitational waves: Waves in a gravitational field traveling at the speed of light mediating the gravitational force.

graviton: A quantum of gravitational energy.

hadrons: Two groups of particles of matter that participate in the strong interaction: baryons with three quarks and mesons with one quark and one antiquark.

half-life: The time required for half of radioactive particles to decay.

halo: The diffuse cloud of particles and dust as well as stars and globular clusters that surround a spiral galaxy.

Heisenberg uncertainty principle: States that it is impossible to measure at the same time the position and momentum of a particle of matter.

helium (He): The second lightest element, consisting of two protons and two neutrons in the nucleus and two electrons in orbit.

Hertzsprung–Russell diagram: A diagram in which the luminosity of stars is plotted against their surface temperature.

horizon: The distance in cosmology beyond which no light signal could yet reach. If the universe has a definite age, the horizon is the age times the speed of light.

Hubble's law or constant (H_0): The relationship between the distance of a galaxy and its velocity of recession. The age of the universe is the inverse of H_0.

hydrogen (H): The lightest atom, consisting of one proton and one electron.

hydrostatic equilibrium: A balance between the gravitational forces and the outward gas and radiation pressure of a star.

inflation: In the classical theory of the big bang, the period between 10^{-35} and 10^{-30} sec when the universe enormously expanded due to a phase transition.

infrared radiation: Electromagnetic radiation of 0.001 to 0.01 cm wavelength. Bodies at room temperature radiate in the infrared band.

infrared slavery: The confinement of all quarks inside hadrons.

isotropy: No dependence on direction.

Jeans mass: The minimum mass for which gravitational attraction can overcome internal pressure and create a gravitationally bound system such as a galaxy.

Kelvin: The temperature scale with absolute zero as 0 K, the equivalent of $-273.15\,°C$. Water freezes at 273.15 K and boils at 373.15 K.

kinetic energy: The energy of motion of a mass m at velocity $v = \frac{1}{2}mv^2$.

leptons: Particles of spin $\frac{1}{2}$ that do not participate in the strong interaction. The electron (e), muon (μ), tau (τ), and neutrino.

light-year: The distance light travels in 1 year or 9.4605 million, million kilometers.

magnetic monopole: A large particle with mass near 10^{-8} gram, so far not discovered with a magnetic pole.

main sequence: The position on the Hertzsprung–Russell diagram of a star when, in its development, thermonuclear reaction starts in the core. Very small stars never arrive on the main sequence.

Maxwell equations: Equations that describe electromagnetic fields. Devised by James Clerk Maxwell.

mesons: Strongly interacting, short-lived particles of matter consisting of one quark and one antiquark.

microwave radiation: Electromagnetic radiation of 0.01 to 10 cm wavelength. Bodies with temperature of a few degrees Kelvin radiate mainly in the microwave band.

muon (μ): An unstable lepton, a cousin to the electron but 207 times heavier.

Any diffusely emitting body.

nebulas: Remnants of supernovas, planetary nebulas, galaxies, and ionized clouds of gas in a galaxy.

neutrino (ν): A massless or near massless (perhaps 10–30 eV) electrically neutral particle. It participates with matter only in weak interactions. There are three varieties: electron-neutrino (ν_e), muon neutrino (ν_μ), and tau-neutrino (ν_τ).

neutron (n): uncharged nuclear particle consisting of three quarks.

neutron star: A remnant core of a large star after supernova explosion with neutrons at nuclear density of 10^{14} g/cm^3.

Glossary

Newton's constant (G): In Newton's gravitational theory, the gravitational force between two masses m_1 and m_2 is their product times the constant divided by the square of their distance. $G = 66.7 \times 10^{-8}$ cm^3/g sec^2 or equivalent.

nucleon: A proton or neutron.

Nucleosynthesis: A period early in the universe when hydrogen and helium gas was formed.

parsec (pc): An astronomical unit of distance where the annual shift (parallax) due to the earth's motion is one second of arc. pc $= 3.0856 \times 10^{13}$ km or 3.2615 light-years. Hubble's constant is given in km/sec per Mpc.

Pauli exclusion: No two particles of the same type can occupy the same quantum state. For example, two electrons cannot occupy the same orbit around the nucleus of an atom.

photon (γ): A massless particle associated with electromagnetic waves.

photosphere: The surface of the sun which is visible at only 6000 K.

Planck energy: $M_p = (\hbar c/G)^{1/2}$ where $\hbar$ = Planck's constant, c = speed of light, and G = Newton's gravitational constant. $M_p = 1.1 \times 10^{19}$ GeV $= 2 \times 10^{-5}$ g in mass units.

Planck's constant (h): The energy of a photon $E_{ph} = hc/\lambda$, where c = speed of light and λ wavelength. $h = 6.6 \times 10^{34}$ watt/second^2. $\hbar = h/2\pi$. It is the basic constant of quantum mechanics.

positron (e^+): A positively charged particle that is the antiparticle of the electron.

positronium: A bound state consisting of an electron and positron.

primeval fireball: The fireball of radiation and particles created at the time of the birth of the universe.

protogalaxy: A gravitationally collapsing cloud of matter and radiation that condenses into stars and forms a galaxy.

proton (p): Positively charged nucleon. The nucleus of hydrogen is a proton.

pulsar: A neutron star that is magnetized, rotating at high speed, and emitting radio pulses.

quantum: It is the quantum theory of electromagnetism.

quantum mechanics: A basic theory wherein elementary particles are either waves or pointlike particles of matter. All energy is quantized.

quarks: The fundamental particles of matter. Carry fraction electrical charge, come in several types (as up, down, strange, charmed, and bottom), and are confined in hadrons.

quasistellar objects or quasars: Small astronomical objects that are usually very distant, emit enormous quantities of radiation, and whose age is close to that of the universe.

radiation era: The time from explosion of the fireball to 534,000 years when radiation was dominant over matter in the universe.

recombination: The combining of atomic nuclei with electrons, forming atoms. In cosmology, the creation of atoms (gas) of helium and hydrogen when the newly created universe's temperature dropped to 3000 K.

red giant: A star of up to 4 solar masses which, after the thermonuclear reaction period, explodes; the star becomes a red giant while the core ends up as a white dwarf or neutron star.

rest energy: The energy of a particle of matter at rest which would be released if everything was transmutated into radiation, in accordance with the Einstein formula $E = mc^2$.

Schwarzschild radius: The event horizon of a black hole from which nothing can escape.

singularity: Zero space where all of the matter and radiation of a black hole are concentrated. Density, temperature, and gravity are infinite.

space-time: Three physical dimensions of space, combined with time, a four-dimensional space-time continuum.

special relativity: A set of equations formulated by Einstein, related to space-time coordinates used by different observers, traveling at different speeds, in a way that the laws of nature appear the same to all and the formulation that the speed of light is the maximum speed in the universe and remains unchanged, regardless of the velocity of the observers.

speed of light (c): The fundamental constant of special relativity. $c = 2,999,729$ km/sec. Only massless particles such as photons and neutrinos can travel at the speed of light.

spin: A basic property of elementary particles describing the state of their rotation. Spin is either an integer or half-integer multiplied by Planck's constant $\hbar$.

spiral galaxy: A galaxy with luminous core and spiral arms of gas, particles, and young stars. Masses range from 10^{10} to 10^{13} $M_\odot$.

strangeness (S): The quantum characteristic assigned to particles that contain strange (s) quarks.

strong interaction: Responsible for keeping quarks inside nucleons and nucleons inside the nucleus of atoms.

supernovas: Large explosions of dead stars in which all mass except the inner core is blown out into space. A supernova produces in a few days as much radiating energy as the sun radiates in 1 billion years.

superstring: A multidimensional space where particles of matter and bosons are replaced with one-dimensional strings.

thermal equilibrium: A state of a system in which all parts are at the same temperature.

ultraviolet radiation: Electromagnetic radiation of 10^{-7} cm to 10^{-5} cm wavelength.

virial theorem: For a star or other massive body, the kinetic energy of particles equals half of the gravitational potential energy $(-)$.

Glossary

virtual particles: Particles to be released from the vacuum of space such as electron–positron pairs which disappear very fast and do not violate the conservation law of energy.

W bosons: The intermediary particles that mediate the weak interaction. Positive (W^+) or negatively charged (W^-).

wavelength (λ): The distance between crests in any kind of wave.

weak interactions: Responsible for the decay of particles such as free neutrons or muons and the β radioactivity decay.

white dwarf: Remnant of a star of maximum mass 1.4 $M_\odot$ with a radius of 1% of the sun and density of 10^5–10^8 g/cm^3.

Z bosons: A neutral boson mediating the neutral current force.

Index

Absorption lines, 125–128
Age of universe, 125–135
Alpha decay, 83–90
Angstrom unit, 353
Angular momentum, 7–10, 12, 13
Annihilation of matter, 1–4, 48–50, 201–206
Antimatter, 201–206
Antineutrino, 11
Antiproton, 51–60
Antiquark, 12, 13, 17–19
Asymptotic freedom, 51–60

b (bottom) quark, 17, 19
Background radiation, 124, 125, 345–350
Baryon number, 13
Baryons, 72, 74, 75, 79, 80
Beta radioactivity, 83–90
Big bang singularity, 121–160
Big bang, Velan model, 219–223
Binary stars, 103–105
Birth of the universe
   Velan, 196–223
   Weinberg, 225–229
Blackbody radiation, 172, 173
Black dwarfs, 293–297
Black holes, 91–96
   Cygnus X-1, 104–106
   defined, 91
   and quasars, 106–109
   Velan theory, 320–328
Boltzmann's constant, 353
Bosons, 83–90, 51–60, 41–50, 91–98
Bubble chamber, 4, 184

c (charm) quark, 17, 19
Carbon, 274–278
Chromodynamics, 51–60
Closed universe, 179–223
Cluster of galaxies, 252–266
Color, 51–60
Compton wavelength, 21–35
Conservation laws, 61–70
   angular momentum, 63, 64
   baryon, 65
   charm, 67
   electrical charge, 64
   energy, 62, 63
   hypercharge, 67
   isotropic spin, 70
   lepton, 65
   momentum, 63
   parity, 70
   spin, 64
   spin J, 68, 69
   strangeness, 66
Corona (sun), 281–286
Cosmic radiation background, 124, 125, 345–350
Cosmic radiation NASA orbiter, 345–351
Cosmological models, 121–160
   Velan model, 179–223
Cosmological theories, 121–160
Cosmology, 1–6
Cosmos, 179–181
Coulomb's Law, 44–46
Creation theory
   Velan, 196–223

Creation theory (*Cont.*)
  Weinberg, 225–229
Critical density, 332–334
Cygnus X-1, 104–106

d (down) quark, 12, 13, 19
Death of stars, 297–316
Decay of particles, 71–82
Deceleration parameter, 332–334
Density fluctuations, 245–250
Density of universe, 332–335
Deuterium, 123, 124
Doppler effect, 125–135

Einstein's relativistic properties of particles, 35–40
Electrical force, 44, 45
Electromagnetic radiation, 31–33, 188–191
Electrons, 7–10, 15, 48–50
Electrovolts, 354
Electroweak force, 113–119
Electroweak unification energy, 113–119
Elementary particles, 7–10, 12, 13, 15, 17–19, 48–50
Elliptical galaxies, 252–266
Energy density of radiation, 172, 222–223
Event horizon, 91–96
Evolution of the universe, 228–244
Exclusion principle (Pauli), 24, 25
Expanding universe, 228–244
Expansion time scales, 329–339

False vacuum, 169, 170
Fate of the universe, 329–344
Fermi constant, 83, 86
Feynman diagrams, 16, 49, 50, 84, 85, 87, 205, 241
Fine structure constant, 41, 42
Finite universe, 179–223
Fireball (Velan), 207–218
Flavor, 12, 13, 17–19
Forces of nature, 1–6
  electromagnetic, 41–50
  gravity, 91–98
  strong, 51–60
  weak, 83–90
  unification, 113–119
"Free lunch" theory of creation, 147–150, 156, 157

Free quarks, 196–223
Friedmann's model of the big bang, 135–138

Galaxies, 245–266
Gamma rays, 188–191
Gamma-ray orbiters, 349–351
Gas clouds, 245–266, 267–280
General relativity, 35–40
Gluons, 51–60
God as the Creator, 140, 141
Gravity, 91–98
Gravitons, 91–98
Great attractor, 330–332
Great walls, 330–332

Hadrons, 71–82
Heisenberg uncertainty principle, 22–24
Helium, 123, 124, 274–278
Herzsprung–Russel diagram, 270–273, 287
Higgs vacuum fields, 176–178
Horizons in black holes, 91–96
Horizons in cosmology, 143, 144
Hubble constant, 128–135
Hydrogen, 274–278
Hyperons, 74, 75

Imaginary time, 157–159
Inflation theory, 145–147
Infrared radiation, 188–191
Initial state of the universe, 179–223
Intergalactic gas, 267–269
Iron, 274–278
Isotropy, 121–123

Jeans mass, 246–250
J/ψ (psi) meson, 77, 80–82

K mesons, 76, 77
Kaon particles, 76, 77

Lambda particles, 74, 75
Light
  cone, 158, 159
  energy, 31–33
  speed, 31–33
  theory, 41–50
  wavelength, 188–191
Lepton number, 65
Leptons, 7–10, 15, 41–50

Luminosity, 33

Magnesium, 224–228
Magnetic moment, 10, 13
Main sequence, 270–273, 287
Matter
 birth, 196–223
 density, 332, 334
Maxwell theory of electricity, 44–46
Mesons, 73, 76–78, 80–82
Microwave background radiation, 124, 125, 345–350
Missing mass, 329, 332
Monopoles, 47, 48
Muon, 14–16

Neutrinos, 11
Neutrons, 51–60
Neutron stars, 316–320
Newton gravitation, 91, 92
Nuclear force
 strong, 51–60
 weak, 83–90
Nucleosynthesis, 236, 239–242

Omega particles, 75
Open strings, 250–257
Origin of the universe
 Velan, 196–223
 Weinberg, 225–229
 See also Creation theory
Oxygen, 274–278

Pair creation, 196–203
Parsec, 357
Particles, 1–6
 creation of, 201–206
Pauli Exclusion Principle, 24, 25
Penrose, Roger, 159, 160
Photons, 41–50
Photosphere (sun), 281–286
Planck theory, 21–35
Planetary nebulae, 276–278
Positrons, 7–10, 15, 48–50
Positronium, 48
 para-, 48–50
 ortho-, 48–50
Primordial fireball (Velan), 207–217
Primordial radiation, 167–172

Prominence (sun), 283
Protogalaxies, 245–255
Protons, 51–60
Pulsars, 316–320

Quantum electrodynamics, QED, 41–50
Quantum gravity, 110–112
Quantum mechanics, 21–35
Quarks
 at creation, 183–185
 b, c, s, and t, 17–19
 d and u, 12, 13, 19
Quasars, 106–109

Radiation, 31–32, 188–191
Radiation-dominated era, 235–242, 244
Rate of expansion, 322–336
Recombination, 242–244
Red giant, 279–294
Red shift, 125–135
Relativity, general theory, 35–40
Rest mass energy, 356
Rho mesons, 73–78
Rotation of stars and galaxies, 245–266, 267, 278

Schwarzschild radius, 91–96
Singularity and the big bang, 121–144
Space, 161–176
 in Velan theory, 181, 182
Special relativity, 93–98
Spectral lines, 188–191
Speed of light, c, 358
Spin of particles, 7–10, 15, 48–50
Spiral galaxies, 252–266
Standard cosmological model, 1–6, 121–160
Stars, 267–278
Strangeness, 66, 67
Strange quarks, s, 17, 19
String theories, 250–257
Strong nuclear force, 51–60
Sun, birth, life, and death, 279–294
Superclusters of galaxies, 330–332
Supergravity, 93–98
Supernova 1, 296, 297
Supernova 2, 296–316
Superstring theory, 150–157

t (top) quark, 17, 19
Tau particle, 14–16

Temperature of the universe from birth to present times, 235–244
Temperature of stars, 267–278
Thermal equilibrium, 267–278
Threshold temperature of particles, 197, 198
Time in relativity, 35–40
Time scale, Hubble, 128–135
Twistor theory, 157–160

u (up) quark, 12, 13, 19
Ultraviolet radiation, 187–191
Universe, Velan model
  birth, 179–223
  expansion, 129, 225–245
  fate, 329–346
  today, 246–328
Uncertainty principle, 22–24
Unification of forces, 113–119

Vacuum of space, 161–176
  in Velan theory, 181, 182, 185

Velan multi-universe cosmos, 179–223
Velan theory of black holes, 320–328
Velocity of stars and galaxies, 128–135
Virtual particles, 161–176
Visible light, 188–191

W bosons, 83–90
Wave–particle duality, 25–31
Wave theory, 25–31
Weak nuclear force, 83–90
Weak interactions, 83–90
Weinberg, Steven, 225–229
White dwarfs, 293–297
Wilson, Robert W., 124, 125
World sheet, 153, 155

X rays, 188–191

Z bosons, 83–90
Zero-point radiation, 170–176

The Universe

Solar mass	$M_\odot$	$= 1.989 \times 10^{33}$ g $= 1.116 \times 10^{57}$ GeV
		$= 1.189 \times 10^{57}$ protons
Radius	R	$= 6.96 \times 10^{10}$ cm
Luminosity	$L_\odot$	$= 3.9 \times 10^{33}$ erg/sec
Milky Way galaxy	Mass M_G	$= 1.8 \times 10^{11} M$
Radius of the Velan fireball	R_{FB}	$= 1.17 \times 10^{14}$ cm
Radius of the universe today	R_0	$= 1.4 \times 10^{28}$ cm
Radius at maximum expansion	R_{ME}	$= 2.52 \times 10^{28}$ cm
Time from start to today (age)		18 billion years
Time from start to maximum expansion		35.5 billion years
Time from start to recontraction		71 billion years
Hubble expansion rate today		17.8 km/sec per million light-years
Critical density	d_{crit}	$= 5.67 \times 10^{-30}$ g/cm^3
Density of fireball	d_{FB}	$= 1.3 \times 10^{15}$ g/cm^3
Density today	d_0	$= 13.54 \times 10^{-30}$ g/cm^3
Density at maximum	d_{max}	$= 2.52 \times 10^{-30}$ g/cm^3
Amount of matter	M_U	$= 5.68 \times 10^{56}$ g
Equivalent in solar masses		$2.86 \times 10^{23} M$
Number of baryons		10^{80}
Number of electrons		10^{80}
Number of photons		10^{89}
Number of neutrinos		10^{89}